BEHAVIOURAL PSYCHOTHERAPY:
Maudsley Pocket Book
of Clinical Management

BEHAVIOURAL PSYCHOTHERAPY: Maudsley Pocket Book of Clinical Management

Isaac M. Marks

With the assistance of:
J. Bird, M. Brown, A. Ghosh, D. Greenberg, F. McCafferey,
R. McDonald, E. Millar, R. Newell, D. Samarasinghe

Bethlem-Maudsley Hospital and
Institute of Psychiatry, London

WRIGHT
Bristol
1986

Published under the Wright imprint by
IOP Publishing Limited
Techno House, Redcliffe Way, Bristol BS1 6NX, England

British Library Cataloguing in Publication Data
Marks, Isaac M.
    Behavioural psychotherapy: Maudsley pocket
book of clinical management.
1. Behaviour therapy
I. Title
616.89′142   RC489.B4

ISBN 0 7236 0875 X

Typset by
BC Typesetting,
51 School Road, Oldland Common, Bristol BS15 6PJ

Printed in Great Britain by
Billing & Sons Limited, Worcester

Preface

This handbook is designed to help clinicians assess and treat patients who are suitable for behavioural psychotherapy and to communicate efficiently with supervisors and referral agents. To this end it provides a practical guide to the principles and methods of behavioural psychotherapy, criteria for selecting suitable cases, measuring their progress, and pithily telling the outcome to supervisors and referring agents. With the aid of simple records of treatment sessions, patients' homework between sessions, and measures of outcome, therapists and their supervisors can easily follow what is happening to the patient.

The main clinical problems addressed in this book are phobic disorders of all kinds (including agoraphobia with panic, and social, specific and illness phobias), obsessive-compulsive disorder, social skills difficulties, sexual dysfunction, sexual deviation, and habit disorders such as stammering, aversions, tics and overeating. For most of these problems the book provides specimen interview guides and letters at assessment, discharge and follow-up. The principles and measures can be readily adapted for clinical management in behavioural medicine for disorders such as psychogenic vomiting, constipation and diarrhoea, hypertension and the prevention of heart disease, although most of these are not specifically addressed.

Those who might find this book useful include trainees in behavioural treatment, be they nurses, doctors, psychologists, social workers or occupational therapists. It is a guide to practical management, and those who wish to probe deeper into theoretical issues in the subject can read appropriate references from the reading list.

Most of the book is relevant to any clinical unit practising behavioural treatment. A few parts are parochial in that they apply only to a particular training course at the Bethlem-Maudsley Hospital, but are included as they might offer ideas useful in other settings.

This handbook is divided into four sections:

1. Introduction. This provides brief general information about behavioural psychotherapy, a behavioural glossary, administration in the Psychological Treatment Unit at the Bethlem-Maudsley Hospital, and suggestions for further reading.

2. Clinical Management. For ease of reference a case example is used in each phase of the treatment process from initial referral to follow-up. At each step this case example illustrates the following: (*a*) Decision flow-chart, (*b*) Definition, (*c*) Interview format, (*d*) Check list, (*e*) Presentation and write-up, (*f*) Letter format with example.

3. Tools for Assessment and Reporting. This section is a guide to evaluation, measurement and related background material, including all forms currently in use in the Unit. A range of specimen letters for various types of clinical problem is also included.

4. Appendices. These describe general administrative issues, training facilities and their use.

The original version of this handbook was compiled in 1977 by Dr Julian Bird as an aid to the course in behavioural psychotherapy for adult neurosis run as JBCNS Course 650 for nurse therapists at the Bethlem-Maudsley Hospital. Since 1977 it has been revised five times in the light of further experience. This is the 6th edition.

Contents

1. INTRODUCTION

1.1 What is behavioural psychotherapy?   1
- 1.1.1   Behavioural psychotherapy as self-help   2
- 1.1.2   For whom does it work?   3
- 1.1.3   Behavioural management principles   5
- 1.1.4   Is there a role for cognitive therapy in behavioural management?   7
- 1.1.5   Behavioural glossary   8

1.2 Measurement and evaluation   12
1.3 Suggestions for further reading   13

2. CLINICAL MANAGEMENT

2.1 Screening   17
- 2.1.1   Flow-chart   17
- 2.1.2   Definition   18
- 2.1.3   Interview format   18
- 2.1.4   Criteria for suitability   18
- 2.1.5   Closed circuit television   19
- 2.1.6   Checklist—before and after screening   19
- 2.1.7   Letter format   20
- 2.1.8   Specimen letters   21

2.2 Assessment
- 2.2.1   Flow-chart   22
- 2.2.2   Definition   23
- 2.2.3   Interview format   23
- 2.2.4   Checklist—before and after assessment   23
- 2.2.5   Specimen letter   24

2.3 Treatment session
- 2.3.1   Flow-chart   26
- 2.3.2   Aim and structure   26
- 2.3.3   Issues affecting outcome   26
- 2.3.4   Checklist—before and after each treatment session   27
- 2.3.5   Write-up   27
- 2.3.6   Sample write-up   28

2.4 Discharge
- 2.4.1   Flow-chart   29
- 2.4.2   General   29
- 2.4.3   Interview format   29
- 2.4.4   Checklist   29
- 2.4.5   Letter format   30
- 2.4.6   Specimen letter   30

2.5 *Follow-up*
   2.5.1   General   31
   2.5.2   Interview format   31
   2.5.3   Checklist   31
   2.5.4   Write-up   31
   2.5.5   Specimen case summary   32
   2.5.6   Letter format   34
   2.5.7   Specimen letters   34
2.6 *Case reporting*
   2.6.1   General   36
   2.6.2   On specific occasions   36
2.7 *Administration of measures*
   2.7.1   General   37
   2.7.2   Selection of measures   37
   2.7.3   Frequency of measures   38
   2.7.4   Administration of measures   38
   2.7.5   Issues with specific measures   38
            Problems and targets, ratings of obsessive-compulsives and sexual attitudes
   2.7.6   Graphs   40
2.8 *Other management issues*
   2.8.1   Ethics   40
   2.8.2   Consultations   41
   2.8.3   Strategy   41

3. TOOLS FOR ASSESSMENT AND REPORTING
3.1 *Measures*
   3.1.1   Data summary   44
   3.1.2   Problems and targets   45
   3.1.3   Work and home management, social and private leisure activities   46
   3.1.4   Fear questionnaire (FQ)   47
   3.1.5   Obsessive-compulsive discomfort, time, handicap   48
   3.1.6   Compulsions checklist—self-rating   49
   3.1.7   Dental pain/fear/anxiety rating scale   50
   3.1.8a   Conventional sexual activity   51
   3.1.8b   Unconventional sexual activity   52
   3.1.9a   Sexual attitudes—normal concepts   53
   3.1.9b   Sexual attitudes—deviant concepts   54
   3.1.9c   Scoring sexual attitudes   55
   3.1.10a   Homework diary   56
   3.1.10b   Sexual homework diary   57
   3.1.11   Social situation questionnaire   58

3.2 Graphs
   3.2.1   Problems/targets   59
   3.2.2   FQ—global, total phobia and dysphoria,
           obsessive-compulsive   60
   3.2.3   Work and home management, social and private
           leisure adjustment   61
   3.2.4   Conventional sexual activity   62
   3.2.5   Unconventional sexual activity   63
   3.2.6   Sexual attitudes—normal concepts   64
   3.2.7   Sexual attitudes—normal and deviant concepts   65

3.3 Interview schedules
   3.3.1   Assessment   66
   3.3.2   Agoraphobia   67
   3.3.3   Overeating and binge-eating   69
   3.3.4   Anger management   70
   3.3.5   Habit disorder   71
   3.3.6   Sexual problems   71
   3.3.7   Sexual dysfunction—management outline   73
   3.3.8   Social skills groups—management hints   73

3.4 Specimen letters
   3.4.1   Spider phobia plus rituals—assessment, discharge
           and follow-up   74
   3.4.2   Phobia of being alone with ectopic heart-beats—
           assessment, discharge and follow-up   77
   3.4.3   Injection phobia—assessment, discharge and
           follow-up   84
   3.4.4   Obsessive-compulsive—assessment, discharge and
           follow-up   91
   3.4.5   Social skills deficit—assessment, discharge and
           follow-up   93
   3.4.6   Sexual dysfunction—assessment, discharge and
           follow-up   96
   3.4.7   Sexual deviation—assessment, re-referral and
           follow-up   98
   3.4.8   Eating disorder—assessment, review, discharge
           and follow-up   101
   3.4.9   Stammer—assessment, discharge and follow-up   105
   3.4.10 Irritable bowel syndrome—assessment, discharge
           and follow-up   108
   3.4.11 Unsuitable—problem unpredictable and causes
           no handicap   113
   3.4.12 Unsuitable—due to psychosis   114

3.5 Data coding
    3.5.1   General   115
    3.5.2   Analyses   115
    3.5.3   Accuracy   115

4. APPENDICES
    4.1 General administrative issues   116
    4.2 File content—Therapist and medical folders   121
    4.3 Training facilities and their use   122
    4.4 Secondment   128

INDEX   131

1. Introduction

1.1 WHAT IS BEHAVIOURAL PSYCHOTHERAPY?

Until about 1970 behavioural psychotherapists drew their main theoretical inspiration from laboratory experiments on animals and in human volunteers with specific fears (mainly students who would not normally have sought professional help. Useful as such work is, it is of limited relevance to severe clinical problems (Marks, 1981). Since then a spate of controlled work with patients produced a clinical discipline that stands in its own right and has weathered unfounded criticisms that behavioural treatment produced symptom substitution and was too mechanistic and superficial. The behavioural approach has successfully helped a wide range of anxiety and other disorders. Behavioural psychotherapy comprises a variety of therapeutic methods that aim to change abnormal behaviour directly rather than by analysing hypothesized conflicts. The problem may be a behavioural deficit (e.g. phobic avoidance, social skills deficit, erectile failure) or excess (compulsive rituals, sexual deviation, tics) which causes severe handicap. The main aim is to alter that behaviour which restricts the patient's social, work, and day-to-day activities, thus improving his quality of life.

For anxiety-related problems like phobic and obsessive-compulsive disorders most behavioural treatments employ some form of exposure to situations which evoke fear and/or rituals.

Exposure involves persuading the patient to enter such situations until their avoidance, rituals and anxiety subside, and is preferably carried out in real life rather than in fantasy (Marks, 1987). This exposure principle is at the core of seemingly different techniques of fear-reduction like desensitization in fantasy and in vivo, flooding in fantasy and in vivo, cognitive rehearsal, modelling and shaping and operant conditioning, paradoxical intention and contextual therapy. All these treatment strategies are ways of exposing the patient to the frightening situation until he or she gets used to it. This process could also be called adaptation, extinction, or habituation.

An important question which remains unanswered is why exposure to noxious stimuli should sometimes produce (sensitize) phobias and at

1

other times reduce (habituate) them. In general, prolonged exposure is more effective than short exposure, hours being more therapeutic than minutes. The few documented instances of experimental sensitization used short exposure durations of a minute or less. Apart from exposure for anxiety-reduction there are also many other behavioural therapeutic strategies to teach new skills and reduce unwanted behaviour. Examples are social skills training, sexual skills training, self-regulation, pacing, prompting, coaching, modelling, contingency management and response cost.
management and response cost.

1.1.1 Behavioural Psychotherapy as Self-help

In recent years increasing emphasis has been given to the self-help aspects of behavioural psychotherapy. Behavioural psychotherapy is not something done to the patient but rather is carried out *with* him. In most cases the therapist's role is that of teacher, coach and monitor while the patient executes his treatment himself. What the patient does between sessions is at least as important as what happens within sessions. This book therefore provides a homework diary all patients are expected to fill in daily (Section 3.1.10). Controlled research has validated this approach. Phobics who were merely asked to follow the self-exposure instructions from the book *Living with Fear* (Marks, 1980) improved substantially to 6-month follow-up, and to the same extent as similar phobics who received exposure instructions either from a psychiatrist or from a computer programmed with the self-help instructions from *Living with Fear* (Ghosh et al., 1986). Results for agoraphobics were similar to those for other phobics.

But behavioural psychotherapy is not simply a matter of telling patients to pull themselves together and use their willpower. If phobics and ritualizers relax themselves *ad infinitum*, that will not reduce their fears and rituals. They improve for the most part only if they enter their feared situation and stop avoiding, escaping and ritualizing, staying with their fear until it subsides, i.e. if they carry out exposure. What the therapist does is to give the patient an appropriate structured framework which makes it easier to complete exposure treatment successfully. It is not motivation *per se* that cures, but doing the right thing.

The same emphasis on self-help applies to the treatment of sexual and other problems. A man with premature ejaculation will only learn to prolong coitus if he and his partner repeatedly practise sexual stimulation to the point just short of emission. An anorgasmic woman has to practise sexual homework involving repeated appropriate stimulation, perhaps first alone, and always later with her partner. Someone who is shy has to carry out social homework tasks of gradually increasing difficulty after he has been shown how to do them. A lad with nocturnal enuresis has every night to set the apparatus that will wake him up as soon as he urinates. A stammerer must every day carry out his exercises speaking during diaphragmatic exhalation in situations of graded difficulty. Always the patient is his own healer, working under the watchful eye of his therapist.

It becomes obvious that patients cannot be treated against their will. Their cooperation is essential, and a skilled therapist must recognize his limitations. Patients who do not want treatment cannot be treated with behavioural psychotherapy (although with children their environment can be manipulated to give them incentives to change—this is much more difficult with adults). But then someone with pneumonia has to swallow his penicillin if the drug is to help him, and the sufferer from chronic renal failure has to carry out an endless appropriate routine if he is to survive.

The beauty of behavioural psychotherapy is that the great majority of suitable patients who carry out their treatment properly improve and stay well as a result of their efforts. Up to 4–7 years follow-up most phobics and ritualizers retain their gains after exposure treatment (Marks, 1986). Few treatments in medicine have had such satisfactory longterm follow-ups. The rewards for the cooperative patient (and therapist) are great. It is always heartening to see years of misery due to 8 hours of handwashing a day, or to being housebound from agoraphobia, and the accompanying family disruption, end in a few weeks as the problem recedes in the course of treatment. Other problems take longer to recede as the opportunities for practice are less frequent, e.g. in sexual and social skills training.

Fig. 1 (*see over*) is a conceptual framework of therapy within which behavioural methods operate.

1.1.2 For Whom does it Work?
Behavioural psychotherapy is now the treatment of choice for selected problems totalling perhaps 13% of all psychiatric outpatients (i.e. half of all neurotics). It is especially beneficial in the management of phobic and obsessive compulsive disorders. The gains made after exposure have lasted up to 7 years after completion of treatment (Marks, 1986).

Indications
Amongst adults the cases most likely to benefit from behavioural psychotherapy include:
— phobic disorders: specific, agora- and social phobias
— obsessive-compulsive disorders, especially rituals

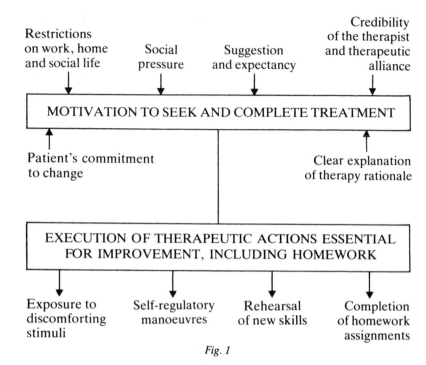

Fig. 1

— sexual dysfunction, e.g. erectile impotence, premature ejaculation, anorgasmia, vaginismus
— sexual deviation (paraphilia), e.g. paedophilia, exhibitionism, fetishism, masochism
— social skills problems
— habit disorders, e.g. hair-pulling, tics, stammering, enuresis, bulimia
— others: nightmares, anger-control, illness behaviour

Not all such cases will respond; the following screening criteria should help to pick out those who will:

i. Therapist and patient agree in defining the problem in terms of observable behaviour.
ii. Current and predictable pattern of the problem behaviours.
iii. Therapist and patient agree on clear behavioural goals.
iv. Patient understands and agrees to the type of treatment offered.
v. Absence of contra-indications, i.e. no severe depression, psychosis or organic illness, no more than 5 mg diazepam or equivalent sedative (or 1 pint of beer or glass of wine) daily. If the patient is taking more of these substances he/she should reduce them gradually to the stated levels before starting behavioural treatment.

Obligations to other patients demand that effort be rationed as follows:
i. Therapy is offered only to suitable patients likely to benefit.
ii. Therapy is stopped after an adequate trial if no major benefit is resulting; 'adequate' refers not just to time spent (norms vary according to diagnosis), but also to the variety and tactics used.

Behavioural psychotherapy has also made inroads in the treatment of childhood behaviour disorders. In addition, under the rubric of behavioural medicine it has begun to have application in the management of hypertension, obesity, irritable bowel syndrome, and other physical disorders. Recently, behavioural family therapy to reduce expressed emotion has proved useful as an adjunct to drug therapy for chronic schizophrenia.

The range of indications for behavioural psychotherapy is gradually extending into new areas. This extension depends upon careful clinical experiments. The indiscriminate use of behavioural treatments in untried areas on unselected patients without proper evaluation leads to much waste of time by both therapist and patient.

1.1.3 Behavioural Management Principles

i. General
There are general principles for all cases, and specific principles for particular types of problems. Within this framework the therapist constructs an individually tailored programme for every patient, selecting from and experimenting with the various approaches (*see* Glossary).

ii. Principles for all cases
1. Negotiation—State the management issues clearly; seek the patient's views directly and state your own frankly. Proceed whenever possible in mutually agreed steps. Aim to make the patient his own therapist.
2. Definitions—Describe all problems, goals, treatments, measures, and outcome clearly and succinctly; use observable behaviour as the key element in all definitions.
3. Goals—Agree goals early; work towards them systematically.
4. Measures—Make them multiple and repetitive; combine broad and narrow scan; involve the patient and others; give feedback.
5. Pragmatism—Make management decisions by results rather than theories. Test 'why' hunches systematically.
6. Practice—Whatever change in behaviour is aimed at (starting, stopping, or modifying), it needs regular practice. Real-life practice

is best—this can be graded; role-play and fantasy are useful adjuncts.
7. Contingencies—Help the patient to examine events that come before, during and after and alter them as needed. Where rewards are used, check that the patient really likes them.

iii. Specific principles
The following are some of the treatment options for these main areas. Their use is individually tailored to the needs of each patient.

1. ANXIETY REDUCTION
Exposure (facing up to the feared situation)
see under: graded exposure (live or in fantasy)
          flooding and implosion
          guided fantasy
          covert sensitization
          response prevention

Anxiety Management
see under: cognitive anxiety management

2. UNWANTED HABITS
Self-control Training
see under: cue exposure
          self-regulation
          thought stopping
          habit control
          contracting

Aversion (pairing the unwanted behaviour with something unpleasant)
see under: classic aversion
          covert sensitization
          response cost

3. DEFECTS OF REPERTOIRE
Skills Training (stepwise modelling and practice of components)
see under: sexual skills training
          social skills training
          modelling
          role play
          shaping
          prompting and pacing

4. MANAGEMENT OF UNUSUAL CASES

As training progresses, trainees undertake the treatment of occasional cases outside the usual remit of behavioural psychotherapy. In these instances the issues of measurement and observable behaviour are critical. A brief trial of treatment may be offered initially with frequent evaluation. Different treatment strategies are introduced sequentially and the use of single case research design is encouraged. New measures specific to the problem may need to be devised and an appropriate literature search should be carried out. Frequent consultation with the supervisor is necessary, particularly regarding the need to determine when an adequate trial of treatment has been completed. An expanded treatment summary detailing the treatment strategies and a possible article for publication maximizes the learning experience, and could eventually lead to the inclusion of the 'experimental' category as a fresh category of patients for treatment by future trainees.

iv. Treatment compliance

Satisfactory treatment outcome depends on the activity of the patient when he or she is not with the therapist. Participation of the patient is essential in behavioural psychotherapy. Expectancy of gains from treatment and desire to change problem behaviour affect a patient's motivation to carry out activities (homework) outside the treatment session. Each therapist, through experience, develops his or her own ideas about what to give as homework assignments and how to increase the likelihood that the patient will complete it. Structuring homework assignments helps both patient and therapist to move purposefully towards a specific treatment target.

A good working relationship between therapist and patient is also important for increasing compliance with often difficult treatment demands. Credibility of the therapist, trust, understanding and respect for the patient's problem promote a successful treatment outcome. Lastly, clear and concise explanation of therapy rationale and strategy helps in alleviating the patient's initial anxiety and builds confidence.

1.1.4 Is there a Role for Cognitive Therapy in Behavioural Treatment?

Many of the manoeuvres included in behavioural psychotherapy concern patients' cognitions (imagery, fantasies and thoughts). This is true for thought-stopping of obsessions, covert sensitization for sexual deviance, and use of guided fantasy exposure where live exposure is not practicable. However, there is as yet relatively little support for the routine use of the cognitive therapy of Beck or the rational emotive therapy of Ellis either for depression or for other problems. Though controlled studies in Philadelphia and in Edinburgh found an advantage in depressives for cognitive therapy over antidepressant drugs, the gains in clinical terms were quite small and at the cost of considerably more therapist time needed per patient, while a study done at the Bethlem-Maudsley Hospital in chronic depressives found no significant benefit over being on a waiting list (Harpin et al., 1982). A recent large well-controlled study in the USA found no advantage in depression of cognitive therapy over interpersonal therapy or over imipramine, compared with drug placebo. Moreover, extensive controlled study has found disappointing results of cognitive methods for social skills deficits, obsessive-compulsive disorder and agoraphobia. Of fully 33 controlled studies of cognitive approaches for these problems, only four yielded convincing benefit from cognitive therapy either on its own or as an adjunct to exposure and skills training (Marks, 1987).

It is quite possible that current research into cognitive therapy will yield more promising results that would justify teaching the approach to trainees interested in routine treatments. That moment has not yet arrived.

1.1.5 Behavioural Glossary

Anxiety: An unpleasant state of apprehension or dread with the expectation that something untoward will occur. It occurs without adequate reason and has autonomic, behavioural and cognitive components.

Anxiety Management (Cognitive): The patient is trained to deliberately alter the things he/she usually says to himself/herself before and during anxiety: new statements are selected and rehearsed; they may amount to *diversion* ('I must make this anxiety worse') or *coping* ('I am going to get through this, even if it is unpleasant: I am unlikely to collapse, but if I do—so what!'). Rewarding self statements can also be useful after an achievement.

Aversion (Avoidance Conditioning): Stimuli that trigger undesirable responses are arranged so that unpleasant events follow, e.g. a paedophile, when he imagines touching a naked child, can snap an elastic band on his wrist, or shock himself or think of something unpleasant like himself being arrested. If this sequence is repeated frequently, the old pleasurable responses die down. To the extent that a patient carries out the aversion himself, this is a form of self-regulation (*see* Self-regulation).

Avoidance: Behaviour which postpones an anxiety-evoking event. It is one of the chief sources of handicap in phobic and obsessive-compulsive disorders.

Behavioural Guidance: Detailed explanation of therapeutic strategy with specific examples, e.g. how to carry out self-directed exposure or self-regulation. It can involve actual demonstration of the method by the therapist and role-rehearsal by the patient.

Contracting: The patient negotiates a formal agreement with another party (e.g. spouse) in which each contracts to make specific changes in their behaviour as desired by the other party. These exchanged behaviours should be simple, positive (active) and frequent.

Covert Sensitization: A form of aversion in which the trigger stimulus and unpleasant consequences are consecutive fantasies. The therapist helps the patient to select and systematically practise these paired fantasies; in addition, the patient is trained to invoke the aversive fantasy when facing real life temptation. Covert sensitization is often combined with cue exposure.

Cue Exposure: Deliberate exposure to 'temptation' (cues, triggers) with the aim of learning to cope in new ways with the urges that follow; this coping may amount to simply waiting till the urge subsides or may involve the deliberate practice of new active alternatives which hopefully are themselves rewarding.

Extinction: Planned ignoring of specific undesirable behaviour. If, in addition, rewards are given for desirable behaviour, this becomes *'differential reinforcement'*.

Fading: Gradual reduction of prompts to the patient to carry out targeted behaviour. Prompts are diminished in frequency or volume, or in other ways, as the patient acquires competence.

Flooding: (*see also* Guided Fantasy; Implosion): Prolonged exposure (live or in fantasy) to the worst feared situation continued until distress reduces. Careful attention is needed to explanations, consent, and medical issues.

Generalization: The spread of a response to a stimulus to other related stimuli.

Graded Exposure: Facing the feared situation in preplanned steps of increasing difficulty ('hierarchy'). This can be live (in vivo) or in fantasy. Progress to a new step is normally delayed until the previous step can be performed with relative ease.

Guided Fantasy: Helping the patient select, build up and rehearse a fantasy. It includes scene setting, 'talking it through', prompting the patient with key words and boosting the fantasy with real objects (e.g. photographs) and choosing an evocative setting for the session.

Guided Mourning: A variant of exposure to reduce morbid grief by bringing the patient into repeated, prolonged contact with cues concerning the deceased, both in imagination and in real life.

Habit Control: Methods to disrupt or adapt habits, such as stuttering, tics, torticollis, cramps, gagging, bed-wetting, etc. These include *'habit-reversal'* (deliberately performing the opposite manoeuvre before and after each habit action); *feedback* (e.g. bell and pad for enuresis); *restructuring* (e.g. new rhythms and pacing for the speech and breathing of stutters); *'massed-practice'* (deliberate repetition to the point of extinction. The setting in which each habit occurs requires equal attention and different techniques (e.g. graded social exposure for stammerers while practising speaking, reward for enuretics the morning after a dry night.

Implosion: Prolonged flooding in fantasy.

Latency: The interval of time between a stimulus and a response to it (e.g. the time the patient takes to produce an aversive image after being asked to do so, as in covert sensitization).

Modelling: Before the patient is asked to do something, someone else demonstrates it to him first. More effective if the demonstrator is seen as resembling the patient, as coping rather than perfect, as gaining reward from the new behaviour, and if the modelled behaviour is in natural short components, followed immediately by the patient doing the same, plus feedback to the patient about how he has done and praise for good performance.

Operant Conditioning (see also Shaping: Token Economy): Arranging the consequences of behaviour by the subject so that desired behaviour earns reward and so increases, while non-desired behaviour leads to reduction of reward or onset of punishment and so decreases. The patient's behaviour is thus arranged to operate rewards and punishments. Also known as instrumental or Skinnerian conditioning, or, with phobics, reinforced practice or successive approximation.

Orgasmic Reconditioning: The patient is taught to switch from deviant to conventional images during masturbation: at the start of training the switch is made close to the point of orgasm: with practice the switch can be made at successively earlier points until finally conventional sexual images are present throughout masturbation and the deviant image is abandoned.

Paradoxical Intention: Reduction of anxiety by encouraging the patient to deliberately try to incur the feared consequence, e.g. a patient afraid of fainting or blushing in public will be asked to deliberately faint or blush in front of other people.

Rehearsal Relief: An exposure method to reduce nightmares by talking and writing about them repeatedly, and giving them a triumphant ending.

Relaxation Training: Induction of general relaxation by teaching the patient to relax his muscles systematically. Each group of muscles is tightened and relaxed systematically with careful feedback. Self-induced relaxation is used as a coping response in the presence of real life anxiety. Relaxation is ineffective in phobic and obsessive-compulsive patients.

Response Cost: A form of aversion (and of self-control training) in which the patient agrees to pay a forfeit, not necessarily monetary, for every failure to carry out a prearranged pattern of behaviour, e.g. every time a cigarette is smoked an agreed sum is put aside to be sent to the patient's 'least favourite' charity.

Response Prevention: A method of prolonging exposure to ritual-evoking cues by asking patients after exposure to such cues to refrain from carrying out the rituals that would usually follow; the urge to perform the rituals then fades.

Satiation: Repeated practice of behaviour till boredom occurs and the behaviour ceases. Uncertain if of lasting efficacy, rarely used. *See* 'massed practice' under 'Habit Control'.

Self-regulation: The patient is taught to monitor, record, evaluate and discriminatively reward or punish his own behaviour and to experiment with new alternatives.

Sexual Skills Training: Stepwise training for a couple to improve sexual behaviour. Components (talking, looking, touching, stimulating, inserting, etc.) are demonstrated (e.g. with books and films) and practised in stages with feedback. Usually there is a ban on attempting more sexual activity than that planned as homework. 'Pause' and 'squeeze' are helpful for improving control over ejaculation, 'tease' for maintaining erection, dilation for reducing vaginismus, and graded masturbation for overcoming anorgasmia.

Shaping ('Successive Approximation'—*see also* Operant Conditioning and Token Economy): A form of operant conditioning in which successive approximations towards the desired new behaviour are rewarded; it is laborious and only used if the desired new behaviour is totally absent from the subject's repertoire and the subject fails to respond to ordinary modelling and skills training.

Social Skills Training: Stepwise training in how to show assertion, warmth, interest, etc. by use of voice, eyes, posture, gesture and key phrases. Components are demonstrated and practised in role play with feedback: this is linked to graded real life practice.

Systematic Desensitization: An elaborate form of graded exposure in fantasy, now obsolete. Brief exposure at each step is paired with the induction of a state that is 'incompatible' with anxiety (usually relaxation) and progress to the next step occurs when the previous step evokes no anxiety.

Taped Audiofeedback: A treatment for obsessive ruminations that is still in its exploratory stages. The patient is asked to prepare an audiotape of their obsessions (e.g. 'I want to have oral sex with him', 'I want to murder her') lasting 45–60 minutes (shorter sequences can be repeated in a continuous loop). He is then asked to carry a walkman tape recorder wherever he goes and to start playing the tape to himself through headphones the moment the ruminations begin, continuing to listen to the tape until the obsessions cease.

Target Behaviour: A concrete, precise component of behaviour which is lacking in the patient's current repertoire and which the patient and the therapist jointly agree to work towards.

Thought Stopping: The patient is asked to invoke an unwanted, obsessive thought in order to learn how to disrupt it: first the therapist makes a loud, sudden noise and gets the patient to do the same: the noise is then progressively lessened, until the patient can achieve the same effect without any outward sign of disturbance. The thought can also be disrupted by snapping an elastic band on the wrist or by self-shocking.

Time-out: Immediately after highly undesirable behaviour the patient is placed somewhere alone for a few minutes without access to rewards.

Token Economy: Arrangement of the patient's milieu so that desired behaviour earns tokens which can be exchanged for rewards. The value of the tokens (amount of rewards they can buy) is periodically adjusted. It is a time-consuming operant approach occasionally used to shape behaviour.

1.2 MEASUREMENT AND EVALUATION

The evaluation of any treatment requires some predetermined criterion for outcome. The simplest is the question 'How do you feel today—better, no different or worse?' This is a crude 'global rating scale'. There is a need for more refined ways to rate patients' problems and treatment outcome.

In physical illness such as hypertension and anaemia, measurement of blood pressure and haemoglobin indicates severity of the problem. For psychological problems a clear and precise definition of the problem behaviour can be measured (e.g. problem and target rating on a 0–8 point scale). This is a fairly reliable and valid way of measuring problem severity. The same scale can and should be used repeatedly during treatment and follow-up to assess progress.

Ratings can be by the patient or an observer (relative, therapist or other), or physiological (e.g. heart rate, skin conductance or penile erection). Physiological measures are valuable for research, but are very rarely necessary in routine clinical practice. The therapist can have most confidence in outcome when there is congruence in the ratings by both patient and observer on a variety of scales.

1.3 SUGGESTIONS FOR FURTHER READING

* Includes self-help sections especially valuable for patients.

Behavioural Psychotherapy: General

Agras W. S. (ed.) (1978) *Behaviour Modification: Principles and Clinical Applications,* 2nd ed., Boston, Little Brown.

Bernstein G. S. (1982) Training behaviour change agents: a conceptual review. *Behav. Ther.* **13**, 1–23.

Fensterheim H. and Glazier H. I. (ed.) (1983) *Behavioural Psychotherapy.* New York, Brunner/Mazel.

Goldstein A. and Foa E. B. (ed.) (1980) *Handbook of Behavioural Intervention—A Clinical Guide.* New York, Wiley.

Marks I. M. (1982) Toward and empirical clinical science: behavioural psychotherapy in the 80's. *Behav. Ther.* **13**, 63–81.

Rimm D. C. and Masters J. C. (1979) *Behaviour Therapy: Techniques and Empirical Findings.* London, Academic Press.

Phobic and Obsessive-compulsive Disorders

Emmelkamp P. M. G. (1982) *Phobic and Obsessive-compulsive Disorders: Theory, Research and Practice.* New York, Plenum Press.

Ghosh A. et al. (1986) Controlled study of self-exposure treatment for phobics. *Br. J. Psychiatry* **148**, in press.

*Marks I. M. (1980) *Living with Fear.* New York, McGraw-Hill.

Marks I. M. (1981) *Cure and Care of Neuroses.* New York, Wiley.

Marks I. M. (1987) *Fears, Phobias and Rituals.* New York, Oxford University Press.

*Mathews A. M., Gelder M. G. and Johnston D. (1981) Agoraphobia: its nature and treatment. London, Guildford Press.

Rachman S. and Hodgson R. J. (1980) *Obsessions and Compulsions.* New York, Prentice-Hall.

Stern R. S. (1978) *Behavioural Techniques.* London, Academic Press.

Social Skills Training

Argyle M. (1983) *The Psychology of Interpersonal Behaviour*, 4th ed. London, Penguin Books.

*Girodo M. (1978) *Shy.* New York, Pocket Books. ISBN: 0671-82599-2.

Liberman R. P. et al. (1975) *Personal Effectiveness.* New York, Research Press.

Stravynski A. et al. (1982) Social skills training in neurotic outpatients. *Arch. Gen. Psychiatry* **39**, 1378–1385.

Anxiety, Depression and Grief

Harpin E. et al. (1982) Cognitive-behavior therapy for chronically depressed patients: a controlled pilot study. *J. Nerv. Ment. Dis.* **170**, 295–301.

Marks, I. M. (1978) Rehearsal relief of a nightmare. *Br. J. Psychiatry* **133**, 461–465.

Mathews A. M. (1981) Anxiety and its management. In: Gaind R. and Hudson B. (ed.). *Current Themes in Psychiatry*, Vol. 3. Oxford, Blackwell.

Mawson D. et al. (1981) Guided mourning for morbid grief: a controlled study. *Br. J. Psychiatry* **138**, 185–193.

Ramsay R. W. (1977) Behavioural approaches to bereavement. *Behav. Res. Ther.* **15**, 131–135.

Whitehead A. and Mathews A. M. (1982) Psychological treatment of depression: A review. *Behav. Res. Ther.* **17**, 495–510.

Marital and Sexual Problems

Bancroft J. H. J. (1983) *Human Sexuality and its Problems*. Edinburgh, Churchill Livingstone.

Cautela J. et al. (1971) Covert sensitization for treatment of sexual deviation. *Psychol. Record* **21**, 37–48.

*Comfort Alex (1972) *Joy of Sex*. New York, Simon & Schuster.

Crowe M. J. (1982) Treatment of marital and sexual problems: a behavioural approach. In: Gorell Barns (ed.). *Family Therapy*, New York, Academic Press.

*Kaplan H. S. (1975) *The Illustrated Manual of Sex Therapy*. London, Souvenir Press.

Hawton K. (1985) *Sex Therapy: A Practical Guide*. Oxford, Oxford Medical Publications.

Linehan K. S. and Rosenthal T. L. (1979) Current behavioural approaches to marital and family therapy. *Adv. Behav. Res. Ther.* **2**, 99–143.

Marks I. M. (1981) A review of behavioural psychotherapy: II. Sexual disorders. *Am. J. Psychiatry* **138**, 750–756.

Masters W. and Johnson V. (1982) *Human Sexual Inadequacy*. Bantam Books.

Riley A. J. and Riley E. J. (1978) Controlled study of directed masturbation for primary orgasmic failure in women. *Br. J. Psychiatry* **133**, 404–409.

Serber M. (1970) Shame aversion therapy. *J. Behav. Ther. Ex. Psych.* **1**, 213–215.

Habit Disorders

Azrin N. H. et al. (1980) Habit reversal versus negative practice—treatment of nervous tics. *Behav. Ther.* **11**, 169–178.

Burns D. and Brady J. P. (1979) Treatment of stuttering. In: Goldstein A. and Foa E. B. (ed.) *Handbook of Behavioural Interventions: A Clinical Guide.* New York, Wiley.

McAuley R. P. (1977) *Child Behaviour Problems: An Empirical Approach to Management.* London, Macmillan.

Moon J. R. and Eisler R. M. (1983) Anger control: An experimental comparison of three behavioural treatments. *Behav. Ther.* **14**, 493–505.

Eating Disorders

Foreyt J. P. et al. (1982) Behavioural treatment of obesity: Results and limitations. *Behav. Ther.* **13**, 153–161.

Hersen M. (1983) Obesity. In: *Outpatient Behaviour Therapy—a Clinical Guide.* New York, Grune & Stratton.

Leitenberg H. et al. (1984) Change during exposure plus response prevention treatment of bulimia nervosa. *Behav. Ther.* **15**, 3–20.

Mahoney M. J. and Jeffrey D. B. (1977) *A Manual of Self-control Procedures for the Overweight.* Washington, DC, American Psychological Association.

Rosen J. C. and Leitenberg H. (1982) Bulimia nervosa—treatment with exposure and response prevention. *Behav. Ther.* **13**, 117–124.

Stuart R. B. (1978) *Act Thin, Stay Thin.* New York, W. W. Norton.

Chronic Schizophrenia and Mental Handicap

Clements J. C. (1979) Goal planning in residential care for the severely mentally handicapped. *Behav. Psychother.* **7**, 1–6.

Fernandez J. (1979) The token economy and beyond. *Ir. J. Psychother.* **2**, 21–41.

Matson J. L. (1980) Behaviour modification for chronic schizophrenics. In: Hersen M., Eisler R. M. and Miller P. M. (ed.). *Progress in Behaviour Modification.* New York, Academic Press.

Wallace C. J. and Lieberman R. P. (1985) Social skills training for schizophrenia: A controlled clinical trial. *Psychiatry Res.* **15**, 239–247.

Behavioural Medicine

Pomerlean O. and Brady J. (1979) *Behavioural Medicine: Theory and Practice.* Baltimore, Williams & Wilkins.

Steptoe A. and Mathews A. M. (1984) *Health Care and Human Behaviour.* London, Academic Press.

Turk C. D., Meichenbaum D. and Myles G. (1983) *Pain and Behavioural Medicine.* New York, Guildford Press.

Childhood Disorders

Douglas J. and Richman N. (1982) *Sleep Management Manual.* Department of Psychological Medicine, Great Ormond Street Children's Hospital, London WC1N 3JH. £1.

Martin H. (1982) *Behavioural Treatment of Problem Children. A Practice Manual.* London, Academic Press.

Yule W. (1977) Behavioural treatment of children's disorders. In: Rutter M. and Hersov L. (ed.). *Children's Disorders.* Oxford, Blackwell.

2. *Clinical Management*

2.1 SCREENING

The point of a brief screening interview is to avoid patients being kept on a waiting list for months before being told they are unsuitable if they turn out to be so. If there is no waiting list, skip screening and proceed straight to assessment (2.2).

2.1.1 Flow-chart

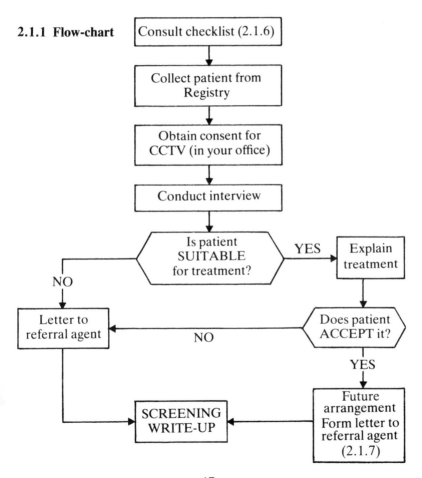

2.1.2 Definition

Screening is a brief interview (20–30 min), to obtain from the patient a clear outline of his presenting problem, to determine his suitability for behavioural psychotherapy, and allow him to make an informed decision about whether or not to accept treatment.

Just before this interview re-read the referral letter and outline the information needed from screening.

2.1.3 Interview Format

1. *Introduction:* Meet and greet the patient, explain the purpose of the interview, and set the time scale.
2. *Obtain a description of the problem:* Ask the patient to outline his problem, then proceed to focused questions about the clinical features of the problem (specific behavioural deficits or excesses, relevant fantasies, physiological sensations), frequency, handicaps, onset, duration, fluctuations, factor(s) that make it worse or better. How would the patient's life change once the problem was eliminated?
3. *Brief mental state examination:* Elicit any contraindications to behavioural psychotherapy (e.g. severe depression, organic brain dysfunction, lack of motivation).
4. *Decide about suitability* on the basis of (2) and (3) above. The outcome could be: (i) suitable for supervised behavioural psycho-therapy, or (ii) suitable for brief behavioural guidance rather than supervised behavioural treatment, or (iii) unsuitable for behavioural treatment. If in doubt discuss with supervisor and peers.
5. *Inform the patient* of the decision. If he is suitable explain what treatment will involve and ask him to nominate a co-therapist where applicable.
6. *Future arrangements:* Set appropriate homework, make a further appointment, write a letter to the referring agent and GP, and write the case up.

2.1.4 Criteria for Suitability for Behavioural Treatment

1. The therapist and patient can agree to define the problem in terms of observable behaviour.
2. This behaviour amounts to a current and predictable pattern.
3. The therapist and patient can agree on clear behavioural goals.
4. There are no contraindications (i.e. no severe depression, psychosis, organic illness, or drug or alcohol dependence).
5. The patient understands and agrees to the type of treatment offered.

6. Behavioural guidance is indicated if the problem is so mild that the patient can overcome it with brief appropriate homework. Systematic supervised behavioural treatment is required if the problem is so complex that the therapist needs to help the patient actually carry out the therapeutic exercises at the start, in addition to the patient's unaided homework.

2.1.5 Closed Circuit Television (CCTV)

Screening interviews are generally conducted under CCTV to facilitate the supervision of trainees and allow them to learn by watching others. Permission from the patient should be sought in your *office* and explained carefully to him *prior* to the interview. No pressure should be exerted on the reluctant patient. Consent should be sought again in the *CCTV room*, prior to commencing the interview. If the interview is being recorded, the patient must be informed and *written* consent obtained, using the form available (Appendix 4.3.3).

2.1.6 Screening Checklist

i. Before screening
1. Registration: All new patients must be registered with the hospital before being seen for a first appointment. The unit secretary shall inform Registry of the time of patients' appointments when these are sent out. If patients present directly to the unit, they should be escorted to Registry for registration before screening.
2. Memorize the patient's name and referral data.
3. Memorize/consider goals of screening, resources and constraints.
4. Have all necessary forms and papers to hand (e.g. CCTV consent, screening pro-forma, rating questionnaires).

ii. After screening
1. Note the outcome: (*a*) in the secretary's master diary (suitable for supervised behavioural psychotherapy or for behavioural guidance only, or unsuitable for behavioural treatment. If one of the first two, note whether the offer of behavioural guidance or treatment was accepted, refused, conditional, or DNA (did not attend)). If suitable, the patient would then be allocated the next available assessment slot, if possible by the screening therapist; (*b*) on the display board giving the name, date of screening, Hospital No. and presenting problem (abbreviated).
2. Letter: findings and decisions (even if DNA (form letter)). In

addition to copy to referral agent and GP, 2 copies for filing and 1 copy each for director and senior registrar.
3. Case flow-chart.

2.1.7 Screening Letter Format
1. Suitable for behavioural guidance or systematic treatment—form letter (*see below*).
2. Unsuitable—letter format:
    a. Identification
    b. State outcome
    c. Brief description of problem
    d. Past history and treatment
    e. Present state
    f. Brief reasons why unsuitable
    g. Management (with any suggestions given)
    h. Future arrangements.
The specimen letter may be helpful when first writing such letters, and contains useful stock phrases (e.g. 'At interview', 'There was/was no evidence'). Later letters will develop a personal style, but should contain the above points.

2.1.8 Specimen Form Letter: Suitable at Screening

(date)

Dear Dr . . . .

Re. . . . . d.o.b. . . . . Hosp. No. . . . .

Address . . . .

Thank you for your referral letter of 12.2.84 about the above patient, who was seen yesterday on behalf of Professor Marks. At initial screening the patient seems suitable for behavioural psychotherapy, but final decision awaits more detailed assessment. This will be carried out on 6.3.84, after which we will write to you in full about our findings.

Yours sincerely

. . . .

Nurse Therapist   Senior Registrar
Professor Marks' Unit

Specimen Letter: Unsuitable at Screening

(date)

Dear Dr . . . .

Re. . . . . d.o.b. . . . . Hosp. No. . . . .

Address . . . .

Thank you for referring this young man, who was seen yesterday for assessment for possible behavioural psychotherapy. His main problem is occasional difficulty in talking to groups, at which times he feels tense and light-headed. This occurs only if the topic discussed is unfamilar or boring to him. The longer he remains silent the more difficult it is for him to break into conversation. He finds it easier to talk to women than to men, and can talk within a group if the subject is of interest and particularly if he is well acquainted with one of the group members. However, he does not avoid social or group activities. His problem has improved over the last three years, and at present he is better than he has ever been.

At interview no deficits were elicited in eye contact, speech or posture, and he was able to assert himself appropriately. As his difficulty in initiating or maintaining conversation in groups is already improving and occurs only very intermittently, he would probably improve still further by extending his repeated practice of taking an active role in groups along the lines he has already begun. I encouraged him to do this and reassured him that his difficulty is shared by many people, especially at his age. He does not need formal social skills training, which is intended for more severe problems. I told him that we would be happy to reassess him should his difficulty become more frequent or handicapping, and that he should seek re-referral through you should that occur.

Yours sincerely

. . . .

Nurse Therapist   Senior Registrar
Professor Marks' Unit

2.2 ASSESSMENT

2.2.1 Flow-chart

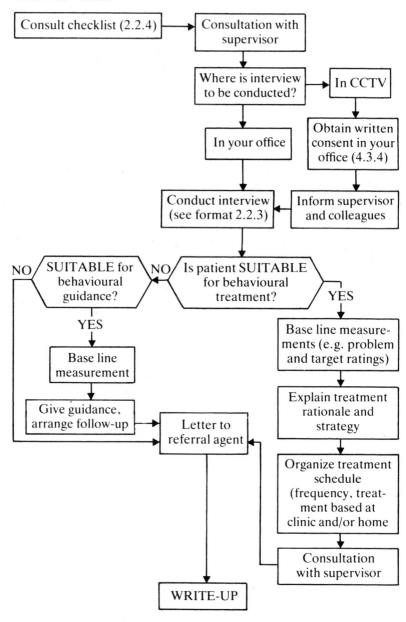

2.2.2 Definition

Detailed assessment (1–2 hours) of a patient's problem is undertaken if he/she has been found suitable for behavioural psychotherapy at screening. The aims of assessment are: (*a*) concise and measurable definition of the problem, (*b*) establishment of treatment targets, (*c*) detailed history of the onset, course and duration of the problem, (*d*) eliciting of information about past difficulties and the social, family and personal history, and (*e*) rating of the problems, targets and associated symptoms (*see* Section 3.1).

2.2.3 Interview Format

1. Introduction: Same as for screening interview (2.1.3(1)).
2. Detailed description of the patient's current problems: Use the same format as at screening, but the information should be more detailed to allow construction of a precise treatment plan.
3. Background information: Past disorders, their treatment and outcome, current living conditions and social activities, previous personality, family and personal history.
4. Mental state examination: Appearance, behaviour, speech, mood, thinking and sensorium. The latter includes orientation, memory (short and long term), concentration (serial 7s), and attention span.
5. Base line measurement: Completion of problem and target rating scales, fear questionnaire, work, home, social and private leisure activities, and other relevant rating instruments depending on the presenting problem (*see* Section 3.1).
6. Explanation of treatment strategy: This should be done with compassion and in plain English, checking that the patient understands each treatment step. Give examples where necessary. Explain the therapeutic rationale—how the problem developed, what is maintaining it, how treatment works and how it would benefit the patient.
7. Future arrangements.

N.B. Consultation with your supervisor before and after the assessment interview is essential in the initial stages of training.

2.2.4 Assessment Checklist

Before assessment

i. Memorize the patient's name, referral and screening data.
ii. Memorize goals of this assessment, resources and limits.
iii. Have all likely forms to hand (e.g., assessment pro forma, measures, diaries, instruction sheets).

After assessment
  i. Write up: do it as soon as possible.
 ii. Next step—what, where, when, who—note in:
        patient's diary
        case folder
        appointments book
        your diary
iii. Amend (*a*) display board
              (*b*) caseflow chart
 iv. Presentation: book with chairperson of next review meeting.
  v. Letter to referral agent.

2.2.5 Specimen Letter: Assessment

(date)

Dear Dr . . . .

Re. . . . . d.o.b. . . . . Hosp. No. . . . .

Address . . . .

Further to my letter of 4.3.83, this patient was seen again yesterday for detailed assessment. She has been frightened of crowds for 14 years; this fear has gradually worsened over the past year. When entering crowds she panics and if she does not escape feels that she will faint. Panic is evoked by large department stores, busy restaurants, including the canteen and coffee room at work, parties, pubs, discos and queueing in busy supermarkets. Currently she avoids entering such situations alone, but is able to do so if accompanied by her husband, mother, a close friend or if exit is easy.

A second problem is fear of talking to people on a one-to-one basis, particularly to those in authority and to an audience; this, too, has worsened over the last year. She now finds it difficult to talk to her supervisor and higher manager at work and avoids them whenever possible. She also panics and fears she will faint in discussions with five or more people, especially at meetings at work, for fear of being the centre of attention. She is on diazepam 2 mg bd and does not abuse alcohol.

Past treatment and investigations for these problems include: an EEG at Queen Mary's Hospital when she was 15—this was normal; attendance at Sutton General Hospital outpatients at age 17 and 19—she had diazepam and clomipramine to no avail, and stopped these as they made her too sleepy; 1980–1981—exposure therapy for 6 months, from a Nurse Therapist at St. George's Hospital, Tooting—she improved about 40% for 2 years.

She lives with her husband in their own home. They have no children. Her problems cause minimal friction in the marriage, her husband being understanding and supportive. She has always been an anxious person, though fairly sociable.

She feels close to her parents. Her father lost his speech during the war and was treated at Banstead Hospital: he has bronchitis and retired from his job as a lorry driver due to this. Her mother is a retired clerk and is in good health. The patient's childhood was enjoyable. She left school at 15 and worked as an accounts clerk for 5 years and as a receptionist for building contractors for 2 years. Four years ago she returned to her previous job as an accounts clerk, remaining there until now. She married 3 years ago, her husband, an accountant, being the same age. Sexual and marital adjustment are satisfactory.

At interview she was neatly dressed and a good informant. Initially she was anxious but this reduced as the interview progressed. She was unhappy about her problems but not depressed and there was no evidence of psychosis. She understands the treatment plan and is keen to have treatment. Her husband was seen separately and gave a history of the problem compatible with hers.

Management of her agoraphobia and social phobia will be on an outpatient basis by graded exposure to the feared situations in real life, starting with the fear of crowds. The patient will be encouraged to take a lead in planning and monitoring her exposure treatment program, and she and her husband have agreed to his involvement as a co-therapist.

We will write again on discharge.

Yours sincerely,

. . . .

Nurse Therapist,   Senior Registrar
Professor Marks' Unit

2.3 TREATMENT SESSION

2.3.1 Flow-chart

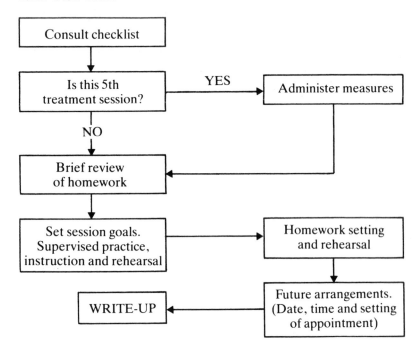

2.3.2 Aim and Structure

Therapists will develop individual styles as confidence and experience grow, but treatment will follow a general plan to ensure progress towards treatment goals at reasonable speed.

1. Administer any necessary forms.
2. Review homework briefly.
3. Negotiate treatment aims with patient.
4. Supervise patients' practice aimed at goals set above. This may be open-ended, the therapist ensuring that practice does not end until significant targets, e.g. anxiety reduction, have been achieved.
5. Review the session (and praise amply for progress made).
6. Negotiate homework and show how to record it.
7. Arrange next appointment.

2.3.3 Issues affecting the Outcome of each Session include:

1. Length and frequency: the longer the session, the more opportunity

there is for anxiety to habituate in phobics and obsessive-compulsives (one and a half hours or more are preferable for them). Intersession intervals of more than a week reduce the chance of closely monitoring patients' homework between sessions, unless there are phone calls in between.
2. Observation and measurement of relevant parameters, e.g. subjective anxiety to monitor progress within the session.
3. Enhancement of motivation to comply with treatment requirements and carry out homework.
4. Flexibility—redefine goals within sessions in the light of the patient's progress so that he can gain maximum benefit from the session.

2.3.4 Treatment Session Checklist

i. *Before each treatment session:*
1. Memorize previous relevant data.
2. Consider goals of this session, resources and constraints.
3. Have all necessary forms to hand.

ii. *After each treatment session:*
1. Next step.
2. Write up the session.

2.3.5 Write-up
Treatment details should be written down immediately after the session, should be brief but informative to others, and should use rating scales as appropriate. The write-up should include:
1. Homework performed.
2. Within-session events.
3. Homework set.
4. Problem areas.
5. Forward planning.
6. Total therapist hours: the total time spent with the patient, whether gathering information, supervising practice or on the telephone. Exclude time spent getting to and from the patient's home or workplace.
7. Supervised practice hours: the total time spent with the patient supervising actual exposure, modelling social skills and other active therapeutic measures. Exclude time spent gathering information or dealing with crises not relevant to the main problems.

2.3.6 Sample Write-up

Patient .... Hosp. No...... Therapist ....

Date 25.2.83   Session No. 3   Session Type*: (E) Croydon shopping centre

SESSION DETAILS: Homework—all tasks
for week completed. Discussion re: diazepam
reduction—reduce from 2 mg bd to 1 mg bd.

	ANXIETY RATINGS		
	Before	During	After
Marks & Spencer—accompanied by therapist 10 minutes	4	3	2
—upstairs alone, 17 minutes	6	4	2
Centre of shopping arcade—twice accompanied, 10 minutes	5	4	2
—alone, 15 minutes	4	3	2
British Home Stores—accompanied, 10 minutes	3	4	2
—alone, 20 minutes	6	3	2

COMMENTS: Tearful at beginning of session, high anticipatory anxiety. Later pleased with her efforts and took lead in deciding treatment tasks (situations likely to evoke high anxiety). No panics during session.
In future treatment sessions and homework—needs to confront more crowded situations, viz, on a Saturday morning.

HOMEWORK:
1. To shop in Croydon for 2 hours on Wednesday and Saturday.
2. To stay in the staff canteen at work for 1 hour once a day.
3. Shop and queue in Sainsbury's for 45 minutes alone, twice a week.

	INVESTMENT	
	Supervised pract. hours	Total therapist hours
This session	1½	2¾
Cumulative total	2½	6¼

NEXT APPOINTMENT:

* Type of session: (C) clinic attendance; (D) domiciliary visit; (E) supervised excursion, e.g. public place; (T) telephone.

2.4 DISCHARGE

2.4.1 Flow-chart

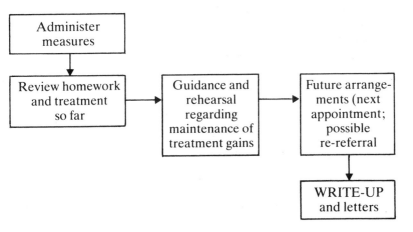

2.4.2 General
A patient is discharged from active treatment when he has improved enough (a drop of at least 4 points in 0–8 rating scales for problems and targets and more if the patient wants to attain that) or if he has completed an adequate trial of treatment with all worthwhile treatment strategies having been tried in vain.

2.4.3 Interview Format
1. Review progress.
2. Administer measures.
3. Identify associated or potential problems.
4. Plan the management of identified problems, e.g. referral to appropriate agencies.
5. Arrange follow-up.

2.4.4 Discharge Checklist
1. Next step.
2. Write-up.
3. Letters.
4. Caseflow chart and display board.
5. Hospital folder content.
6. Data coding.
7. Book presentation for case review with patient in attendance.

2.4.5 Discharge Letter Format

'Further to my letter of (date) this patient has now been discharged from active treatment.' Then short paragraphs on:
1. Main problems treated (definitions only).
2. Main problems not treated (brief).
3. Type of treatment (avoid jargon).
4. Outcome: what improved, got worse, and did not change (give examples). 'We will write again after 1-month follow-up.'

2.4.6 Specimen Letter: Discharge

(date)

Dear Dr ....

Re. .... d.o.b. 5.5.1954 Hosp. No........

Address ....

Further to my letter of 22.3.83 this patient has now been discharged from treatment. The main problems treated were: (*a*) inability to enter crowded situations like large department stores, restaurants, parties and queueing in busy supermarkets, for fear of fainting; (*b*) avoidance of talking to people on a one-to-one basis (especially those in authority) and speaking in front of an audience, for fear of fainting and being the centre of attention.

She had 10 sessions of exposure treatment involving gradually entering the feared situations and remaining in them until her anxiety reduced. Her husband was actively involved in treatment as a co-therapist.

The agoraphobia and social phobia problems are much improved. The patient now regularly goes alone into stores, restaurants and supermarkets, enjoys parties, and can speak to people in authority or in front of an audience with only minimal anxiety; previously these situations would have evoked fear. Overall anxiety is improved and her fear is much less but still occurs on occasions—mainly in the canteen at work, but she feels able to cope with these. Her diazepam was reduced from 2 mg bd to 0·5 mg prn. Her husband confirms her report of great improvement.

She has been advised to continue her exposure exercises and to stay in the fear-evoking situation until she feels easier in them. We will write again after one month follow-up.

Yours sincerely

....

Nurse Therapist
Professor Marks' Unit

2.5 FOLLOW-UP

2.5.1 General

Follow-up interviews are conducted to the same general plan as treatment sessions. If all is going well with the patient, follow-up gives the opportunity to reinforce gains made and offer advice for further consolidation, whilst the therapist fades out contact with the patient. Problems are identified and rectified by further treatment during the follow-up sessions, guidance as to 'homework', or a further period of supervised therapy (booster sessions).

2.5.2 Interview Format

1. Administer measures.
2. Review progress since discharge and 'homework'.
3. Identify actual or potential problems.
4. Set goals based on problems identified.
5. Supervise practice, or give behavioural guidance, or offer further treatment sessions.
6. Review sessions.
7. Set homework.
8. Next appointment.

Though the standard frequency of follow-up in the unit is 1, 3 and 6 months, the therapist should offer additional follow-up if the patient's condition warrants it. The standard follow-ups are a minimum. Discharge itself is part of a gradual fading of therapist contact, and 'booster sessions' should be offered when necessary.

2.5.3 Follow-up Checklist

At follow-up, the therapist checks:
1. Next step.
2. Write-up.
3. Letters.
4. Caseflow chart.
5. Hospital folder content.
6. Data coding.
7. Presentation booked (including presentations with patient in attendance).
8. Case summary.

2.5.4 Write-up

Write-up of follow-up sessions follows the format for treatment sessions. The case summary should be written at 1 month follow-up, on standard sheets and should be no longer than 2 sides of an A4 sheet.

A copy goes in each patient's file, plus one each to the Director and to the Senior Registrar.
1. Identification and treatment dates.
2. Assessment—follows format of assessment letter, but abbreviated.
3. Treatment—type, principles.
   Special features (domiciliary, co-therapist, role play).
   Sessions—number, frequency, duration, total therapist time.
4. Outcome—what got better, worse, remained unchanged—main scores.
5. Follow-up—changes since discharge.

2.5.5 Case Summary

MAUDSLEY HOSPITAL   CONSULTANT:
PROF. I. MARKS

BEHAVIOURAL PSYCHOTHERAPY

Patient .... Hosp. No. .... d.o.b. 5.5.1954

Address ....

Therapist Karen Smith   Treatment began: 8.3.83.   Ended: 26.7.83.

Referral: Dr   (GP)

Main Problem:
1. Severe anxiety amounting to panic when entering crowded places unaccompanied. Avoids large department stores, busy restaurants including canteen and coffee room at work, parties, pubs, discos and queueing in busy supermarkets. Onset was sudden in 1969 after returning from holiday in Spain, where there was a typhoid scare. Convinced she had typhoid, teased at swimming baths by attendant about this. After swimming she went shopping, panicked in large department store and left immediately. The fear is eased when accompanied in the feared situations and by diazepam 2 mg.
2. Fear of talking to people on a one-to-one basis, particularly those in authority, and also speaking in front of an audience of five people or more. Avoidance of talking to immediate superior at work and attending meetings. Onset gradual; worsened over last year. Eased by diazepam 2 mg.
Alcohol: Socially (2 glasses wine weekly). *Medication:* Diazepam 2 mg bd.
Previous Treatment: Psychiatric 1971—Queen Mary's Hospital; E.E.G. 1973—Sutton General Hospital; prescribed diazepam and clomipramine. 1975—Outpatient Sutton General Hospital; same medication and psychotherapy. 1980–81—St. George's Hospital as outpatient for behavioural psychotherapy, 40% improvement, for 2 years. 1982—one session of private psychotherapy.
   Medical: 4 yrs inpatient Romford and Billericay Hospitals for operation on brain clot. 10 yrs—appendectomy, Southend General Hospital.

Social: Lives in own house with husband. Accounts clerk. Lots of friends; now rarely socialises due to problem.

Forensic: Nil.

Family History: Father aged 64, has bronchitis; lost speech during war, treated at Banstead Hospital. Mother aged 58, in good health. Patient an only child. No known family psychiatric history.

Personal History: Happy childhood, no problems. School aged 5–15; average scholar, had friends. On leaving school became accounts clerk; then receptionist for building contractors: last 5 yrs account clerk, at same firm.

Sexual History: Menarche age 14; informed by mother. One sexual partner beside husband. Many casual boyfriends. Coitus now on alternate days and is satisfactory for both. Happily married 3 yrs to an accountant of same age.

Mental State: Neatly dressed. Good informant and co-operative. Unhappy about problem but not depressed. No evidence of psychosis or disorientation.

Treatment: As an outpatient. Graded exposure in real life used for both problems. 10 sessions: no missed appointments; total therapist hrs 20, supervised hrs 12.

Outcome: She is now able to enter all feared situations with minimal anxiety. Diazepam reduced to 0·5 mg prn. Gains in Problem 1 maintained to one month follow-up. Reports a return of difficulties in Problem B at follow-up, but continues to practise her exposure exercises.

	Pre	Post	1-mfu	3-mfu	6-mfu
Problem A Inability to enter crowded places on my own, leading to avoidance of such situations, particularly large department stores, restaurants, parties and queueing in busy supermarkets, for fear of fainting.	6	2	3	6	3
Target A1 To queue in a busy supermarket on my own, once a week, for 1 hour.	8	2	3	7	4
Target A2 To walk around a large department store alone, twice a week, for 2 hours.	8	2	3	6	3
Target A3 To sit in the middle of the canteen at work, once a day, for ½ hour.	8	6	8	8	4
Problem B Avoidance of talking to people on a one-to-one basis, especially those in authority, and speaking in front of an audience, for fear of fainting and being the centre of attention.	8	3	6	8	5
Target B1 To sit in a room and talk to five or more people for 1 hour once per day.	8	4	8	8	6
Target B2 To talk to my immediate supervisor once per day for 15 minutes	6	2	7	8	4

2.5.6 Letter Format
1. Identification and follow-up date and point.
2. Reminder of problem treated (very brief).
3. Current state of patient, mental status and problem—emphasize behaviour.
4. Changes for better or worse, residual difficulties.
5. Date of next appointment.

2.5.7 Specimen Letter: 1-Month Follow-up

(date)

Dear Dr ....

Re. .... d.o.b. 5.5.1954. Hosp. No...........

Address ....

Further to my letter of 2.8.83, this patient was unable to attend two appointments arranged for her 1-month follow-up. I subsequently posted the relevant self-rating forms to her, which she has completed and returned. Her agoraphobia remains improved in that she is able to enter large department stores and queue in busy supermarkets. However, her social anxieties at work have returned. They concern talking to her immediate supervisor and in front of a group of people (five or more), and sitting in the staff canteen.

She is continuing to practise her exposure exercises in these situations and has mapped out a detailed self-exposure treatment program for herself. I have arranged to see her on 15.11.83 for 3-month follow-up; if she then still has problems in these areas, I will arrange 'booster' sessions of behavioural treatment.

We will inform you of her progress.

Yours sincerely,

Nurse Therapist   Senior Registrar
Professor Marks' Unit

Specimen Letter: 3-Month's Follow-up

Dear Dr ....

Re. .... d.o.b. 5.5.1954. Hosp. No...........

Address ....

Further to my letter of 11.10.83, this patient was seen for 3-month follow-up on 21.12.83 and again on 3.1.84 after missing several appointments.

She is again unable to enter crowded situations on her own or talk to people on a one-to-one basis, for fear of fainting or panicking. At both interviews she was accompanied by her husband, whose report confirmed hers. After discharge she did not continue her exposure exercises whilst at work, where her difficulties gradually recurred and then spread to department stores, supermarkets and restaurants. She has now resigned from her job after much discussion with her husband. Both felt that she would have done this even if her problems had not returned. She plans to get over her problem, find part-time employment and to start a family in the near future.

I have arranged with her to resume active behavioural treatment. Both she and her husband feel she would benefit from this, and she will be able to devote more time to the treatment programme, since she is no longer working.

As you are aware, she is on clomipramine 10 mg tds and diazepam 2 mg tds. She ceased taking clomipramine as she complained of 'sickness'. Currently she takes diazepam 2–4 mg before going out. I have advised her to reduce this and to avoid any diazepam before treatment sessions or before carrying out exposure-homework tasks, as it could reduce their therapeutic effect.

We will keep you informed of her progress.

Yours sincerely, etc.

Specimen Letter: 6-Month's Follow-up

(date)

Dear Dr ....

Re. .... d.o.b. 5.5.1954. Hosp. No. ....

Address ....

Further to my letter of 10.1.84, this patient failed to attend regular 'booster sessions' or her 6-month follow-up appointment. Consequently, follow-up was conducted by sending her relevant forms by post. She completed and returned these together with a letter about her progress concerning her agoraphobia and social phobia.

She has not been practising her exposure exercises as advised and has made no further gains. She has left her job and does not intend to return to work. Thus she no longer encounters the social situations at work that she fears, and she is mostly accompanied by her husband in other crowded public places. She wrote that she believed she would improve once they have settled in their new environment, and would then recommence her exposure exercises on a self-help basis. Whether this will actually happen remains to be seen.

We will see her for 1-year follow-up.

Yours sincerely

Nurse Therapist   Senior Registrar
Professor Marks' Unit

2.6 CASE REPORTING

2.6.1 General

Case review meetings are held twice weekly during intensive training—once for outpatients and once for inpatients. All current inpatients are reviewed every week, but outpatients are presented only at selected points in their treatment programme, namely at assessment, after every five sessions, at problem points, at discharge and at follow-up. Because of pressure of business at case reviews, the presentations have to be brief (*10 min* for assessment, *2–5 min* for others), and yet succinct and easily followed. A standard format is therefore followed as below; at discharge and follow-up data coding sheets are also presented. Medical case notes and case folder should be at hand during presentations at assessment.

2.6.2 On Specific Occasions

1. *At assessment*
   Follow the assessment write-up format (2.2.5)
2. *After every five sessions*
   Identification
   Referral
   Problem and target definitions
   Sessions done—number? frequency? duration? type?
   Changes—in treated problems (show graphs – 3.2)
           —in untreated problems
   Revised predictions
3. *At problem points*
   Identification
   Referral
   Problem and target definitions
   Sessions done—number? frequency? duration? type?
   Changes—in treated problems (show graphs)
           —in untreated problems
   The management problem—definition
                         —options available
                         —preferred option with reasons
4. *At discharge and follow-up* (minimum 1, 3 and 6 months—more often when necessary)
   Identification
   Referral
   Read discharge/follow-up letter (2.4.6, 2.5.7)
   Show graphs
   Show data summary (and case summary, if applicable)
   Show data coding

2.7 ADMINISTRATION OF MEASURES

2.7.1 General

Measurement is an important aspect of therapy, providing valuable feedback to the patient, therapist and supervisor. Whether things are going well or badly, good measurement enhances awareness of this and increases the likelihood of appropriate action being taken.

Initial measures focus the client on his current state, and link assessment to treatment. Grasping the point of measurement can aid the patient to see his own problem more objectively and record home-work more accurately. Measurement is an opportunity for teamwork between patient and therapist in defining goals.

To be realistic, measurements should be relevant, carefully and consistently explained and applied, and regularly repeated. They should involve reports from relatives or other observers.

2.7.2 Selection of Measures

The following measures are applied to all cases:
Problems
Targets
Fear and anxiety–depression questionnaire
Work/home/social leisure/private leisure
The following additional measures are required for specific categories:

Obsessive-compulsive	*Stammer*
time/discomfort/handicap	dysfluencies per minute
compulsion check list	words per minute
Sex dysfunction	*Sex deviation*
conventional attitudes	deviant attitudes
conventional urges/acts/satisfaction	deviant urges/acts/satisfaction
sexual activity diary	conventional attitudes
	conventional urges/acts/ satisfaction
	sexual activity diary

During treatment and homework, measure the pleasure accompanying deviant imagery/scenes and aversiveness of aversive imagery. Time taken to obtain images need not be measured (3.3.6).

Other clinical problems

Individual new measures should be constructed for any patient where the standard measures are not a realistic guide to his state. In such cases devise with the patient a way of scoring the intensity and/or frequency of whatever event treatment should decrease or increase—

e.g. marital rows, hair pulling, nasty feelings, number of eating binges, drinks taken, frequency and extent of bed-wetting, etc.

When in doubt about adequacy of the measures then insist on a *daily diary* about specific behaviour.

2.7.3 Frequency of Measures

The minimum frequency for all measures, once selected, is:
    before treatment (at least once; preferably more)
    during treatment (at least every 5 sessions)
    at discharge
    at 1-, 3- and 6-month follow-up
Clearly some measures have to be made frequently (e.g. number of tics per hour).

2.7.4 Administration of Measures

Take care regarding:
1. Explanation: read through the explanatory notes of each form with the patient, eliciting feedback point by point. Give examples.
2. Avoiding contamination: make your own ratings before you see the patient's self ratings. Ensure that the patient cannot see his previous ratings, either from session to session, or intrassession.
3. Consistency: administer measures, as far as possible, with standard explanations from patient to patient and session to session. Rate at a uniform time on each occasion (preferably at the start) so as to avoid the tendency to score low following a single successful session.

2.7.5 Issues with Specific Measures

1. *Problems and targets* (*see* 3.1.2): Devising and rating problems and targets is the patient's introduction to measurement. This exercise is introduced with: 'At this stage we need to decide what our priorities are and how we are going to measure progress as we go along. This will help us to give each other quick and clear feedback and will avoid unnecessary muddle or upset or wasted time . . .'.

'*Problems*' are introduced roughly as follows:
'The first step is to describe the main problems in a nutshell in one or two short sentences to help us measure how severe they are . . .'.

The final problem definitions should:
—wherever possible, be in the patient's own words

—clearly state the patient's original complaints
—concern observable behaviour, either as a deficit or an excess, including speech, thoughts or feelings
—indicate the frequency of this behaviour
—indicate any central idea or feeling associated with it
e.g. 'Avoidance of being alone, leaving the house, using public transport or shopping, due to fear of collapsing'.

'Targets' are introduced with: 'The next step is to pinpoint some target tasks that you will aim to do regularly by the end of treatment. Targets should be practical and useful things to do; they should be very specific so we may need several for each problem'.

The final target definitions should be:
—an active (positive) item of behaviour. Behavioural targets are always preferable to 'feelings' targets
—useful and desirable to the patient
—able to reflect progress on one of the problems
—defined closely (duration, frequency, circumstances)
—an item of behaviour which, wherever possible, involves the patient in learning skills which will generalise to problem areas not specifically targetted
e.g. 'To travel by tube to work in Covent Garden, daily for at least one hour'.

2. *Ratings of obsessive-compulsives (see 3.1.5)*: This can become an obsessive pursuit by the patient unless the therapist insists on time limits, approximations, first time ratings. This in itself is an experience of therapy for the obsessive patient, emphasizes consistency on the part of the therapist, and demonstrates the rationale of treatment to the patient.

3. *Sexual attitudes (see 3.1.9)*: To convert the 'X's into scores, use the following key for both heterosexual and deviant scales:

sexless	6543210	sexy
makes me calm	0123456	makes me anxious
cruel	6543210	kind
erotic	0123456	frigid
pleasant	0123456	unpleasant
makes me tense	6543210	makes me relaxed

For attitudes to sexually deviant concepts, low scores denote pathology. In order to follow the convention that 'down' on the graph indicates improvement, for the graphs of sexually deviant concepts the polarity of scoring on the vertical axis is reversed (*see* 3.2.7).

The blank headers for the sexually deviant concepts and activities measures should be filled in after negotiation with the client and follow the same conventions as problem statements.

2.7.6 Graphs

Transcribing clinical measures onto graphs assists efficient monitoring of cases. It makes for quick feedback to patient, therapist, and supervisor alike, and tracks the impact of any changed circumstances in a way that other records do not. When looking at case graphs, lessons are often learnt and running adjustments made that would otherwise have been missed.

Graphs should be constructed when necessary in a standard way:

Thus (*see* illustration for numbered points, section 3.2).
1. Clear title
2. Clear key
3. Scores always shown on vertical axis (higher being more severe)
4. All scores of equal value are at equal height above base
5. Time always shown on the horizontal axis
6. All scores done at the same time point are at equal distance from the margin
7. Indicate breaks in time scale
8. Indicate incidental events (e.g. start of drugs) with arrows on the time axis.

2.8 OTHER MANAGEMENT ISSUES

2.8.1 Ethics

If you act in good faith and are open and honest with the patient, you will find few problems. When in doubt, ask yourself the question: 'Would I like things to be done this way, if I (or my wife or mother or best friend) were the patient?'

Problems that do arise with the kind of patients seen in this unit usually concern:
1. *Confidentiality:* all clinical data are confidential; the potential for causing social distress or damage of any given bit of information will not always be obvious—it depends on the particular circumstances of the patient. The rule is that information can only be shared with those professionally concerned with this patient's care—for anyone else, including other professionals and relatives, you must seek explicit, preferably written, permission from the patient. There are special issues with videotapes, and additional rules to follow (*see* Appendix).
2. *Consent:* for potentially distressing or dangerous procedures, make sure that the patient not only signs the written consent form (*see* Appendix), but that he/she definitely understands the nature of the procedure, the reasons for it, the risks, and the alternatives, before

going ahead. Typical examples are CCTV interviews and use of explicit sexual material. If there is doubt about the patient's understanding then do not persist.
3. *Sex:* showing sexually explicit material to the patient requires written consent (*see* Appendix): whenever possible, involve the patient's sexual partner in such sessions.
4. *No smoking policy:* Therapists can be potent models for patients and thus should not smoke in their presence.

2.8.2 Consultations

You are expected to consult supervisors over doubts concerning selection, management and medical issues. The initiative should come from you.

Before consulting you should prepare:
1. A clear statement of the clinical problem.
2. A short list of options as to how to proceed.
3. Your preferred option, with reasons.

2.8.3 Strategy

i. *General*

Any treatment has known indications, goals, limitations, and requirements for success. The aim is efficient deployment of time and skills to treat as many suitable patients as possible.

ii. *Decision flow*

Successful management usually follows a series of crucial steps in which each new step depends on successful achievement of the preceding step. Naturally there are exceptions—steps that look crucial, but turn out not to be, or look irrelevant yet turn out to be crucial; nevertheless it is useful to try and make strategic and tactical decisions with a stepwise sequential map or 'flow-chart' in mind.

Crucial steps may be of small 'tactics' (such as the patient grasping the use of clinical ratings) or large (such as completing an adequate assessment interview). Each step carries criteria (sometimes not explicit) for successful achievement. The attempt to pass each step generates one of three decisions:
1. 'Achieved' (proceed to next step).
2. 'Doubtful' (worth another try with new tactics).
3. 'Impossible' (no realistic hope of progress).

iii. Strategic issues are those that should be considered in advance of sessions:

Before taking on the patient:
1. Note criteria of selection—what makes a case suitable or unsuitable? (2.8.2)
2. Resources—what kind of treatment and travelling time, knowledge, equipment, personnel might be needed? What deficiencies can/cannot be made up? Is a suitable local co-therapist necessary and available, especially if the patient lives far away? Are there some otherwise suitable cases that should not be taken on because of these deficiencies?

Before each session with the patient note:
1. Goals of the meeting—what new information, feelings, attitude or behaviour do I hope to achieve in this meeting?
2. Resources—what time, knowledge, equipment, people, setting would be optimal to reach these goals? Can such resources be provided? What are the alternatives?
3. Follow-up: Detailed measurement of progress at follow-up is made at one, three and six months and one year after discharge. Where needed, follow-up interviews may be more frequent. Follow-up is also a time for reassessment of the patient's current status and for booster treatment where necessary.

iv. Note-taking during interviews: This will sometimes be essential, e.g. at assessment. It is equally crucial that it would not interfere with information flow or the relationship. Three manoeuvres can help: tell the patient of the need to write, use personal shorthand, and stick only to hard data ('pegs' to hang your write-up on).

3. *Tools for Assessment and Reporting*

3.1 MEASURES

3.1.1 Data Summary

3.1.2 Problems and Targets

3.1.3 Work and Home Management, Social and Private Leisure Activities

3.1.4 Fear Questionnaire

3.1.5 Obsessive-compulsive Discomfort, Time, Handicap

3.1.6 Compulsion Checklist

3.1.7 Dental Pain/Fear/Anxiety

3.1.8a Conventional Sexual Activity

3.1.8b Unconventional Sexual Activity

3.1.9a Sexual Attitudes (Heterosexual)

3.1.9b Sexual Attitudes (Deviant)

3.1.9c Scoring Sexual Attitudes

3.1.10a Homework Diary

3.1.10b Instructions for Recording Sexual Homework

3.1.11 Social Situations Questionnaire

3.2 GRAPHS

3.2.1 Problems/Targets

3.2.2 Fear Questionnaire—Global, Total Phobia and Dysphoria, Obsessive-compulsive

3.2.3 Work and Home Management, Social and Private Leisure Adjustment

3.2.4 Conventional Sexual Activity

3.2.5 Unconventional Sexual Activity

3.2.6 Sexual Attitudes (Normal Concepts)

3.2.7 Sexual Attitudes (Normal and Deviant Concepts)

3.1.1 Data Summary

BEHAVIOURAL PSYCHOTHERAPY   MAUDSLEY HOSPITAL

Patient .... Hosp. No........ Occupation (or spouse's) .... d.o.b. .... 19..
Address .... tel. .... Relative/friend ....   age [ ]
Ref. from .... tel. .... tel. M1.F2
G.P. .... tel. .... Therapist IP1.OP2
                                                                                 Mar1.sng2
                                                                                 div./sep.3, wid.4

	Screen	Pre.	Post.	1m	3m	6m
DIAGNOSIS 1. 2.	Date →					
TREATMENTS USED: 1. Main 2. Extra						
PROBLEM A duration [ ] yrs — Self						
— Ther.						
TARGET A1 — Self						
— Ther.						
TARGET A2 — Self						
— Ther.						
PROBLEM B duration [ ] yrs — Self						
— Ther.						
TARGET B1 — Self						
— Ther.						
TARGET B2 — Self						
— Ther.						
UNTREATED 1. — Self						
PROBLEMS 2. — Self						
Main phobia						
Fear questionnaire: — Self						
General anxiety — Self						
Compul. checklist — Self						
Work — Self						
Home management — Self						
Social leisure — Self						
Private leisure — Self						

INVESTMENT
total   [ ] sessions
domiciliary   [ ] sessions
missed   [ ] appointments
total ther. time [ ] hours
ther.-supervised [ ] hours
family co-therapy [ ] yes 1, no 2

FOLLOW-UP ARRANGEMENTS

DISPOSAL CATEGORY [ ]
1 = completed treatment
2 = lapsed before session 2
3 = lapsed between session 3 & 6
4 = other (specify) ....

3.1.2 Problems and Targets

Patient ....   Hosp. No. ....   Therapist ....

Date ....

Problems

		Pre.	Mid.	Post.	1 mfu	3 mfu	6 mfu
A	Self						
	Therapist						

		Pre.	Mid.	Post.	1 mfu	3 mfu	6 mfu
B	Self						
	Therapist						

'This problem upsets me and/or interferes with my normal activities'

0	1	2	3	4	5	6	7	8
does not		slightly/ sometimes		definitely/ often		markedly/ very often		very severely/ continuously

Targets

		Pre.	Mid.	Post.	1 mfu	3 mfu	6 mfu
A1	Self						
	Therapist						
A2	Self						
	Therapist						
B1	Self						
	Therapist						
B2	Self						
	Therapist						

'My progress towards achieving each target regularly without difficulty'

	0	2	4	6	8
DISCOMFORT/ BEHAVIOUR	none/ complete success	slight/ 75% success	definite/ 50% success	marked/ 25% success	very severe/ no success

3.1.3 Work and Home Management, Social and Private Leisure Activities

Hosp. No. . . . .

Patient . . . . Therapist . . . .

	Pre. date	Mid. date	Post. date	1 mfu date	3 mfu date	6 mfu date

Work

'Because of my problems my ability to work is impaired:'

```
0      1      2      3      4      5      6      7      8
not at      slightly      definitely      markedly      very severely
all                                                      I cannot work
```

Self

Ther.

Home Management   (cleaning, tidying, shopping, cooking, looking after home or children, paying bills)

```
0      1      2      3      4      5      6      7      8
not at      slightly      definitely      markedly      very severely
all                                                      I cannot do it
```

Self

Ther.

Social Leisure Activities   (with other people, e.g. parties, pubs, clubs, outings, visits, dating, home entertainment)

```
0      1      2      3      4      5      6      7      8
not at      slightly      definitely      markedly      very severely
all                                                      I never do these
```

Self

Ther.

Private Leisure Activities   (done alone, e.g., reading, gardening, collecting, sewing, walking alone)

```
0      1      2      3      4      5      6      7      8
not at      slightly      definitely      markedly      very severely
all                                                      I never do these
```

Self

Ther.

3.1.4 Fear Questionnaire

Name ....   Age ....   Sex .....   Date ....

Pre./Post./1m/ 3m/6m

Choose a number from the scale below to show how much you would avoid each of the situations listed below because of fear or other unpleasant feelings. Then write the number you choose in the box opposite each situation.

0——1——2——3——4——5——6——7——8

would not ... *slightly* ... *definitely* ... *markedly* ... *always*

avoid it ... *avoid it* ... *avoid it* ... *avoid it* ... *avoid it*

1. Main phobia you want treated (describe in your own words) – ....
2. Injections or minor surgery ....
3. Eating or drinking with other people ....
4. Hospitals ....
5. Travelling alone by bus or coach ....
6. Walking alone in busy streets ....
7. Being watched or stared at ....
8. Going into crowded shops ....
9. Talking to people in authority ....
10. Sight of blood ....
11. Being criticized ....
12. Going alone far from home ....
13. Thought of injury or illness ....
14. Speaking or acting to an audience ....
15. Large open spaces ....
16. Going to the dentist ....
17. Other situations (describe) ....

Ag + Bl + Soc = Total

2–16

Now choose a number from the scale below to show how much you are troubled by each problem listed, and write the number in the box opposite.

0——1——2——3——4——5——6——7——8

hardly ... *slightly* ... *definitely* ... *markedly* ... *very severely*

at all ... *troublesome* ... *troublesome* ... *troublesome* ... *troublesome*

18. Feeling miserable or depressed ....
19. Feeling irritable or angry ....
20. Constant tension wherever I happen to be ....
21. Sudden surges of panic regardless of where I am ....
22. Upsetting thoughts coming into your mind ....
23. Other feelings (describe) ....

Total

How would you rate the present state of your main problem on the scale below?

0——1——2——3——4——5——6——7——8

phobias ... *slightly* ... *definitely* ... *markedly* ... *very severely*

absent ... *disturbing/* ... *disturbing/* ... *disturbing/* ... *disturbing/*

not really ... *disabling* ... *disabling* ... *disabling*

disabling

Please circle one number between 0 and 8

3.1.5 Obsessive-compulsive Discomfort, Time, Handicap

PROBLEM A

PROBLEM B

Patient .... Therapist ....
Hosp. No. ....

	Pre. date 19....	Post. date 19....	1 mfu date 19....	3 mfu date 19....	6 mfu date 19....
	A ☐ B ☐	A ☐ B ☐	A ☐ B ☐	A ☐ B ☐	A ☐ B ☐
	A ☐ B ☐	A ☐ B ☐	A ☐ B ☐	A ☐ B ☐	A ☐ B ☐
	A ☐ B ☐	A ☐ B ☐	A ☐ B ☐	A ☐ B ☐	A ☐ B ☐

'Discomfort when unable to perform this activity is:'

0	1	2	3	4	5	6	7	8
absent		slight		definite		marked		extreme

rate here ↗↖

'Total time each day this activity takes me:'

0	1	2	3	4	5	6	7	8
0–5 min	5–15 min	15–45 min	45–75 min	1¼–2 hr	2–3 hr	3–5 hr	5–8 hr	8 hr +

rate here ↗↖

'Handicap: Compared to most people this activity in my case is:'

0	1	2	3	4	5	6	7	8
no different		twice as lengthy or frequent/ slightly avoided		3 times as lengthy or frequent/ definitely avoided		4 times as lengthy or frequent/ markedly avoided		5 times as lengthy or totally avoided

rate here ↗↖

3.1.6 Compulsion Checklist

Name . . . . Hosp. No. . . . . . . Date . . . . 19 . . .

INSTRUCTIONS: The following are a list of activities which people with your kind of problem sometimes have difficulty with. Please answer each question by putting a tick under the appropriate number

0 – 'I have no problems with activity—takes me about the same time as an average person. I do not need to repeat it or avoid it.'

1 – 'This activity takes me about twice *as long as most people, or I have to repeat it* twice *or tend to avoid it.'*

2 – 'This activity takes me about three *times as long as most people, or I have to repeat it* three *or more times, or I usually* avoid it.'

3 – 'I am unable to complete or attempt activity.'

				ACTIVITY
				Having a bath or shower
				Washing hands and face
				Care of hair (e.g. washing, combing, brushing)
				Brushing teeth
				Dressing and undressing
				Using toilet to urinate
				Using toilet to defaecate
				Touching people or being touched
				Handling waste or waste bins
				Washing clothes
				Washing dishes
				Handling or cooking food
				Cleaning the house
				Keeping things tidy
				Bed making
				Cleaning shoes
				Touching door handles
				Touching your genitals, petting or sexual intercourse
				Visiting a hospital
				Switching lights and taps on or off
				Locking or closing doors or windows
				Using electrical appliances (e.g. heaters)
				Doing arithmetic or accounts
				Getting to work
				Doing your work
				Writing
				Form filling
				Posting letters
				Reading
				Walking down the street
				Travelling by bus, train or car
				Looking after children
				Eating in restaurants
				Going to public toilets
				Keeping appointments
				Throwing things away
				Buying things in shops
TOTAL				
				Other (fill in)

3.1.7 Dental Pain/Fear/Anxiety

Name ....
Hosp. No. .... Date........

Select the number on the scale below that corresponds
with your feelings during the following dental
procedures and write it in the first column below.

| 0 | 1 | 2 | 3 | 4 | 5 | 6 | 7 | 8 |

no distress   some   definite   strong   extreme
at all   uneasiness   distress   distress   distress

Dental Procedures	Pre.	Mid.	Post.	1 mfu	3 mfu
a. Receiving an appointment					
b. Making an appointment					
c. Visiting for an examination					
d. Having an X-ray					
e. Having an impression for dentures					
f. Having a scaling					
g. Having a tooth drilled					
h. Having an injection (local anaesthetic)					
i. Having a tooth drilled under local anaesthetic					
j. Having a tooth drilled under general anaesthetic					
k. Having a tooth drilled under sedation					
l. Having a tooth extracted under local anaesthetic					
m. Having a tooth extracted under general anaesthetic					
n. Any other procedure					

ADDITIONAL ITEMS   Date ___ ___ ___ ___

Is fear of pain, of gagging, or of both?
Pain/gag   Age acquired
Related experiences
Family history of dental phobia
General stress factors/allergies
Drugs: tablets, alcohol, cigarettes
Catarrh, sinus, throat, lungs, indigestion
Sensitivity of teeth and of mouth
Associated phobias/rituals
Anxiety/precipitants

3.1.8a Conventional Sexual Activity

Hosp. No. . . . . Rated by self/partner . . . .

Patient . . . .   Therapist . . . .

Date	Pre. . . . . . 19 . .	Post. . . . . 19 . .	1 mfu . . . . 19 . .	3 mfu . . . . 19 . .	6 mfu . . . . 19 . .
	☐	☐	☐	☐	☐
	☐	☐	☐	☐	☐
	☐	☐	☐	☐	☐

'I have sexual intercourse:'

0	1	2	3	4	5	6	7	8
not at all		once a month		once a week		2–3 times a week		daily

'I have sex play with my partner:'
(petting, foreplay)

0	1	2	3	4	5	6	7	8
not at all, less than once a month		once a month		once a week		2–3 times a week		daily

'For me enjoyment during these sexual activities is usually:'

0	1	2	3	4	5	6	7	8
absent		slight		moderate		marked, but without orgasm		marked, usually to orgasm

3.1.8b Unconventional Sexual Activity

Hosp. No. .... self-rating

Patient .... Therapist ....

ACTIVITY A

ACTIVITY B

Date	Pre. 19	Post. 19	1 mfu 19	3 mfu 19	6 mfu 19
	A ☐ ☐ B	A ☐ ☐ B	A ☐ ☐ B	A ☐ ☐ B	A ☐ ☐ B
	A ☐ ☐ B	A ☐ ☐ B	A ☐ ☐ B	A ☐ ☐ B	A ☐ ☐ B
	A ☐ ☐ B	A ☐ ☐ B	A ☐ ☐ B	A ☐ ☐ B	A ☐ ☐ B

'I get urges to do this:'   rate here ↘ ↗

0	1	2	3	4	5	6	7	8
less than once a month		once a month		once a week		once a day		3 times daily or more

'I masturbate or have sex while thinking about this:'   rate here ↘ ↗

0	1	2	3	4	5	6	7	8
less than once a month		once a month		once a week		once a day		3 times daily or more

'I carry out this activity:'   rate here ↘ ↗

0	1	2	3	4	5	6	7	8
less than twice a year		every 3 months		once a month		once a week		daily

3.1.9a Sexual Attitudes (Heterosexual)

Patient . . . .   Therapist . . . .

Date . . . . pre/post/1 m/3 m/6 m

Look at each *concept* written in capitals below (e.g. YOUR CURRENT SEXUAL PARTNER). Then look at each *scale* below it (e.g. sexless—sexy) to see if the concept has a connection with one or other end of the scale. If there is *no* connection, put a cross in the middle of the scale, like this: sexless __:__:_×_:__:__ sexy.
If you think your CURRENT SEXUAL PARTNER is *extremely* sexy, place your cross like this: sexless __:__:__:__:__:_× sexy,
if *moderately* sexy, like this: sexless __:__:__:__:_×_:__ sexy,
if only *slightly* sexy, like this: sexless __:__:__:_×_:__:__ sexy, etc.
Remember to put your cross on a line, like this __:__:_×_:__
not like this __:_×_:__:__

YOUR CURRENT SEXUAL PARTNER

sexless __ : __ : __ : __ : __ : __	sexy
makes me calm __ : __ : __ : __ : __ : __	makes me anxious
cruel __ : __ : __ : __ : __ : __	kind
erotic __ : __ : __ : __ : __ : __	frigid
pleasant __ : __ : __ : __ : __ : __	unpleasant
makes me tense __ : __ : __ : __ : __ : __	makes me relaxed

E S A

NORMAL SEXUAL INTERCOURSE

sexless __ : __ : __ : __ : __ : __	sexy
makes me calm __ : __ : __ : __ : __ : __	makes me anxious
cruel __ : __ : __ : __ : __ : __	kind
erotic __ : __ : __ : __ : __ : __	frigid
pleasant __ : __ : __ : __ : __ : __	unpleasant
makes me tense __ : __ : __ : __ : __ : __	makes me relaxed

E S A

AN ATTRACTIVE MEMBER OF THE OPPOSITE SEX

sexless __ : __ : __ : __ : __ : __	sexy
makes me calm __ : __ : __ : __ : __ : __	makes me anxious
cruel __ : __ : __ : __ : __ : __	kind
erotic __ : __ : __ : __ : __ : __	frigid
pleasant __ : __ : __ : __ : __ : __	unpleasant
makes me tense __ : __ : __ : __ : __ : __	makes me relaxed

E S A
(< = reversed
polarity)

3.1.9b   Sexual Attitudes (Deviant)

(*Note:* FILL IN APPROPRIATE DEVIANT CONCEPTS AT A, B AND C)
Patient . . . .   Therapist . . . .

Date pre./post./1m/3m/6m

Look at each *concept* written in capitals below (e.g. EXPOSING MY GENITALS TO
ATTRACTIVE FEMALES). Then look at each *scale* below it (e.g. sexless—sexy) to
see if the concept has a connection with one or other end of the scale. If there is
no connection, put a cross in the middle of the scale, like this:
sexless __:__:__:×:__:__:__ sexy.

If you think EXPOSING MY GENITALS TO ATTRACTIVE FEMALES is
extremely sexy, place your cross like this: sexless __:__:__:__:__:__:× sexy,
if *moderately* sexy, like this: sexless __:__:__:__:__:×:__ sexy,
if only *slightly* sexy, like this: sexless __:__:__:×:__:__ sexy.
Remember to put your cross on a line, like this: __:×:__:__
not like this: __:__×__:__:__

e.g. EXPOSING MY GENITALS TO ATTRACTIVE FEMALES (A)

sexless __ : __ : __ : __ : __ : __ : __ sexy
makes me calm __ : __ : __ : __ : __ : __ : __ makes me anxious
cruel __ : __ : __ : __ : __ : __ : __ kind
erotic __ : __ : __ : __ : __ : __ : __ frigid
pleasant __ : __ : __ : __ : __ : __ : __ unpleasant
makes me tense __ : __ : __ : __ : __ : __ : __ makes me relaxed

E S A

•e.g. MASTURBATING TO FANTASIES OF EXPOSING MY
GENITALS (B)

sexless __ : __ : __ : __ : __ : __ : __ sexy
makes me calm __ : __ : __ : __ : __ : __ : __ makes me anxious
cruel __ : __ : __ : __ : __ : __ : __ kind
erotic __ : __ : __ : __ : __ : __ : __ frigid
pleasant __ : __ : __ : __ : __ : __ : __ unpleasant
makes me tense __ : __ : __ : __ : __ : __ : __ makes me relaxed

E S A

(FILL IN RELEVANT DEVIANT CONCEPT) (C)

sexless __ : __ : __ : __ : __ : __ : __ sexy
makes me calm __ : __ : __ : __ : __ : __ : __ makes me anxious
cruel __ : __ : __ : __ : __ : __ : __ kind
erotic __ : __ : __ : __ : __ : __ : __ frigid
pleasant __ : __ : __ : __ : __ : __ : __ unpleasant
makes me tense __ : __ : __ : __ : __ : __ : __ makes me relaxed

E S A
(< = reversed
polarity)

3.1.9c Scoring Sexual Attitudes

Each scale is scored from 0 (sexy, makes me calm, kind, erotic, etc.) to 6 (sexless, makes me anxious, cruel, frigid, etc.). Read 0 from the left to 6 on the right unless there is < in the scoring box in which case score 0 from the right to 6 on the left.

Score range is 0–12 for Evaluation, for Sex and for Anxiety factors (2 scales each).

3.1.10a Homework Diary

Week commencing: ....

Name ....

Goals for the week
1. ....
2. ....
3. ....
4. ....

Anxiety scale:

0	2	4	6	8
no anxiety	slight anxiety	definite anxiety	marked anxiety	panic

Session		Goal No.	Task Performed	Anxiety			Comments incl. coping tactics	Co-signed
Date	Began	Ended			Before	During	After		

3.1.10b Instructions for Recording Sexual Homework

Instructions for Recording Sexual Homework

Could you please record your sexual activities using the key below.
Each symbol denotes a different aspect of lovemaking. The symbols
are to be written along the time line which shows 5-minute intervals.
Your record should begin at 0 minutes; the times you record obviously
cannot be exact, only approximate.

$\sim\sim$	= coitus	o	= woman's sexual arousal
——	= nongenital foreplay, caressing	Ø	= woman's orgasm
m	= manual stimulation of genitals	v	= vibrator
or	= oral stimulation of genitals	mc	= manual stimulation of clitoris
E	= full erection	mv	= manual stimulation of vagina
e	= partial or failed erection	mp	= manual stimulation of penis
↓	= ejaculation		

Activities carried out by the man *and* the woman are written *on* the line;
Activities carried out by the *man only* are written *above* the line;
Activities carried out by the *woman only* are written *below* the line;

Date	Start Time	\multicolumn{13}{c}{Minutes after start}
		0	5	10	15	20	25	30	35	40	45	50	55	60
………	………													
………	………													
………	………													
………	………													
………	………													
………	………													
………	………													
………	………													
………	………													
………	………													
………	………													
………	………													
………	………													
………	………													
………	………													
………	………													
………	………													
………	………													
………	………													
………	………													
………	………													

3.1.11 Social Situations Questionnaire

Name .... Hosp. No. .... Date .... 19 ....

This questionnaire concerns how you get on in social situations (being with other people, talking to them, etc.).

Please rate the discomfort you experience in the situations listed, using the following scale:

0	1	2	3	4
no discomfort	*slight discomfort*	*moderate discomfort*	*great discomfort*	*I avoid this situation*

1. Walking down the street ......
2. Going into shops ....
3. Going on public transport .....
4. Going into pubs ....
5. Going to parties ....
6. Mixing with people at work ....
7. Making friends of your own age.
8. Going out with someone of the opposite sex ....
9. Being with a group of the same sex and roughly the same age as you ....
10. Being with a group of both men and women the same age as you
11. Being with a group of the opposite sex the same age as you ....
12. Entertaining people in your home, lodgings, etc. ....
13. Going into restaurants or cafes .
14. Going to dances, dance halls or discotheques ....
15. Being with older people ......
16. Being with younger people ....
17. Going into a room full of people.

18. Meeting strangers...........
19. Being with people you don't know very well ....
20. Being with friends ....
21. Making the first move in starting up a friendship..............
22. Making ordinary decisions affecting others (e.g. what to do together in the evening) ....
23. Being with only one other person, rather than a group ....
24. Getting to know people in depth
25. Taking the initiative in keeping a conversation going ....
26. Looking at people directly in the eyes ....
27. Disagreeing with what other people are saying and putting forward your own views ....
28. People standing or sitting very close to you ....
29. Talking about yourself and your feelings in conversation ....
30. People looking at you ....

Total

3.2 GRAPHS

3.2.1 Problems/Targets

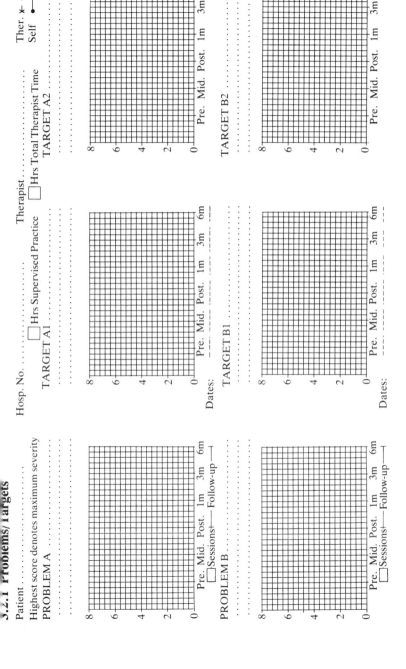

3.2.2 Fear Questionnaire—Global, Total Phobia and Dysphoria, Obsessive-compulsive

Patient . . . .   Hosp. No. . . . .   Therapist . . . .

Highest score denotes maximum severity

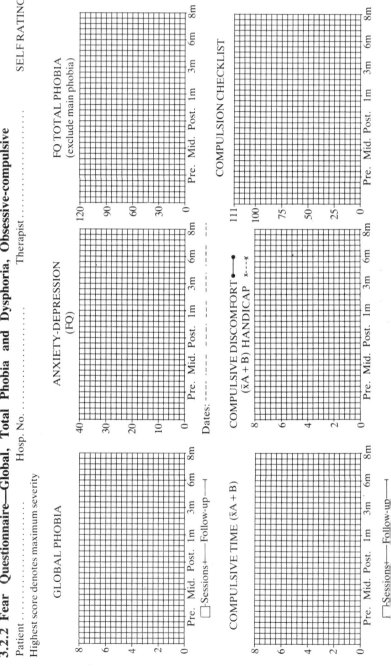

3.2.3 Work and Home Management, Social and Private Leisure Adjustment

Patient.................... Hosp. No...................... Therapist...................

Highest score denotes maximum severity

Self ●——●
Ther. ✶------✶

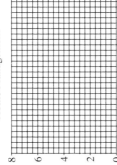

Home Management

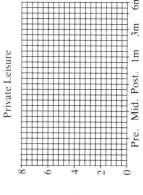

Private Leisure

Dates:------------

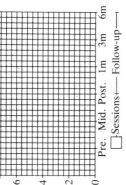

Work

|-Sessions-|——Follow-up—|

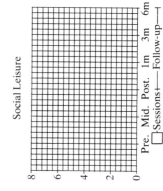

Social Leisure

|-Sessions-|——Follow-up—|

3.2.4 Conventional Sexual Activity

Patient . . . .

Hosp. No. . . . .

Therapist . . . .

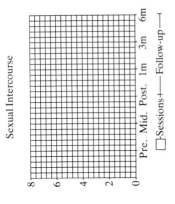

Sexual Intercourse

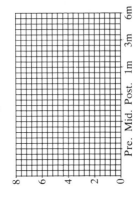

Sex Play With Partner

Self ●——————

Partner ✗- - - - - -✗

Enjoyment of Sexual Activity

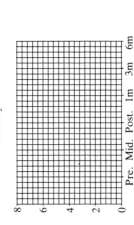

3.2.5 Unconventional Sexual Activity

[Self Rating]

Patient . . . .

Hosp. No. . . . .

Therapist . . . .

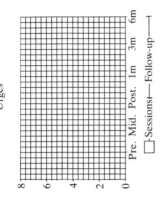

Urges

Acts

☐ Sessions ⊢— Follow-up —⊣

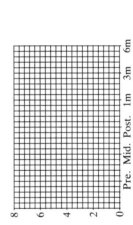

Masturbation
(to Deviant Fantasy)

3.2.6 Sexual Attitudes (Normal Concepts)

Therapist . . . .

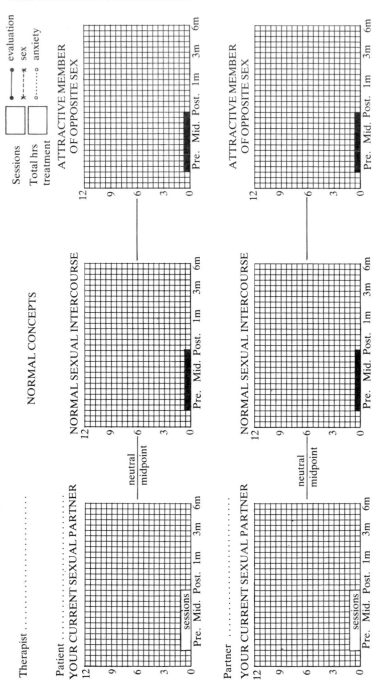

3.2.7 Sexual Attitudes (Normal and Deviant Concepts)

Patient ....

Therapist ....

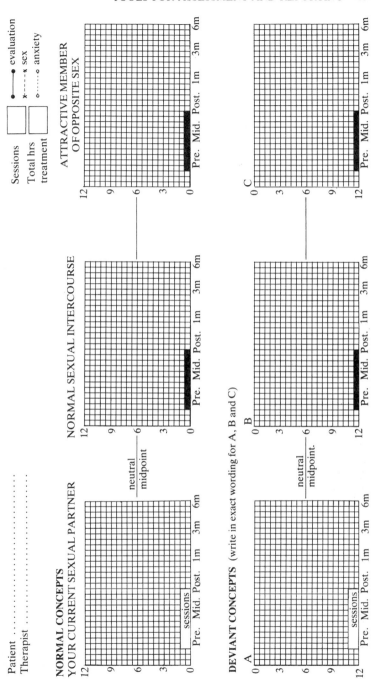

3.3 INTERVIEW SCHEDULES

3.3.1 Assessment

General

This should be done very soon after the assessment interview(s), occupy 2–3 sides of quarto, use headings, and cover the areas below in the sequence listed:

Identification: name, age, home area, married/single, job.

Referral: by whom (inside or outside Maudsley), for what (immediate precipitant).

Main problem(s) (detailed): Typical features: duration—circumstances at onset—fluctuations—what makes it better or worse—impact on self and others—medicines and alcohol (what? how often? how much? for how long?).

Past disorders (very brief): This problem: treatment—dates—places.
Other psychiatric: treatments—dates—places.
Other medical: treatments—dates—places.
Forensic history: charges—court appearances—probation orders—dates.

Social conditions (brief): Home—where? what? with whom? social links?
Work—where? what? with whom? social links? flexible hours?
Leisure—where? what? with whom? social links?

Personality (brief): (character/temperament—get third party comments).

Family history (brief): Parents: age—health: death—date—cause: work: character.
Siblings—number/order (e.g. patient 2nd of 5)—ages—links.
Psychiatric history in family: treatments—dates—places (very brief).

Personal history (brief): Childhood: where born/bred? early symptoms—home atmosphere.
School: duration (ages)—exams—friends—absences.
Further education: duration—exams—friends.
Work: current job: type—place—duration—friends.
　　　longest job: type—place—duration—friends.
　　　other jobs: number—type—duration—dismissals (very brief).
　　　skills/plans (very brief).
Sex (use separate expanded section for sex problems).
　　　ages of puberty, masturbation, sex play, coitus.
　　　partnerships—length and nature.
　　　current activity—partner(s)—habits—frequency—satisfaction.

Marriage: spouse—age—occupation—joint activities.
　　　　　　duration of marriage—plans for future.
　　　　　　past separations—divorces.
Children: ages—genders—problems (very brief).
Mental state: General: dress—manner—style of talk.
Mood: consistent up or down—sleep—appetite—energy—guilt—
　　　suicide—anxiety.
Reality contact: strange beliefs—experiences—thoughts—attitude to
　　　　　　problem and treatment.
Orientation: date—place.
Memory: recent events.
Formulation: This is a brief account of your conclusions based on the
information above, followed by your management plans.

　It must include:
1. A brief 2- or 3-line summary of the main problem, e.g. 'This is a
   35-year-old housewife who has been unable to leave her home
   unaccompanied for the past two years.'
2. Handicaps arising from the problem(s) and factors making it worse
   or better.
3. Details of other problems.
4. Management: —the chosen plan, note special tactics, e.g. hospital-
   　　　　　　ization, co-therapy, role play, homework, home
   　　　　　　treatment.
   　　　　　　—Alternative plans.
5. Likely therapeutic investment (no. of treatment hours needed from
   therapist) and prognosis—what will improve? what will not? how
   many sessions over how long.
6. For unusual problems give literature reference concerning
   treatment.

3.3.2 Agoraphobia

1. What are your main fears (list in order of severity)?
2. When did these fears first begin to trouble you (note date for each
   fear and when avoidance of each situation began)?
3. What feelings do you experience in these situations?
4. What was your worst experience?
5. Have you ever had a panic? Can you describe it to me? What did
   you do?
6. Were there any special circumstances in your life when your
   phobia(s) began?
7. Have you ever received treatment for your phobias: what did this
   consist of?
8. Is your travelling restricted in any way?

9. Do you avoid travelling by bus (car, tube or surface train, plane, ship) or going into streets, shops, theatres, cinemas?
10. How do you feel when you are left alone at home?
11. Do you ever take to your bed because of anxiety?
12. If left alone do you: (*a*) try to call in a friend (how often)? (*b*) telephone your husband (wife, friends, etc.)?
13. Do members of your family have to spend any of their time with you because of your phobias? Do they help with shopping or taking the children to school?
14. Do you take anything with you when you leave the house (e.g. tablets, smelling-salts, sunglasses, umbrella)?

Relevant questions from the following can be put to the patient as appropriate:
Open/closed spaces
15. Do you dislike open spaces (e.g. fields) or enclosed spaces?
16. Do you avoid going into lifts or elevators? tunnels? rooms without windows? How many flights of stairs would you climb rather than take a lift?
17. Can you visit your hairdresser? How do you feel there?

Crowds
18. Do you dislike busy or empty deserted streets?
19. Could you go into a crowded football match? Trafalgar Square?
20. Do you avoid going into cinemas, theatres, churches, concerts?
21. If not, do you sit in any special place, e.g. back or aisle seats?
22. Could you go into a crowded shop?

Heights
23. Do heights bother you?
24. To what floor of a building could you ascend without discomfort?
25. How many steps of a ladder could you climb?
26. Do bridges bother you?

Social situations
27. Could you start talking to a complete stranger?
28. Does sitting round a table at home or in a restaurant trouble you? Is it easier to eat with friends (with strangers or alone)?
29. Do you dislike people looking at you? Could you sit in a bus or train and face the passengers opposite you without discomfort?
30. Do you avoid going to parties?
31. Could you give a short speech to friends?

Other fears
32. Vomiting. Animals. Surgery. Blood. Illness. Being left alone.

Associated symptoms
33. Do you suffer from depression? (note mood: energy: appetite: anhedonia: libido: guilt: attitude to future: suicidal ideas).
34. Do you ever have feelings of unreality or that the world is unreal?
35. Do you have any difficulties in your sex life? women: do you achieve orgasm? men: do you achieve an erection and orgasm?
36. Do you get compulsions to check things, e.g. to make sure that gas taps and plugs are turned off?
37. Are you particular about cleanliness in the home? or worried about catching and spreading disease by contact with dirty objects?
38. Do thoughts keep running around in your head which cause you to repeat actions unnecessarily?

Miscellaneous
39. Has anyone in your family had fears like yours?
40. Have you ever discussed your condition with others who had a similar complaint?

3.3.3 Overeating and Binge-eating

Assessment
1. Types of food eaten.
2. Cues that lead to eating.
3. Where/when food eaten.
4. Associated feelings before and after eating.
5. Induced vomiting/use of laxatives: fasting/dieting: now/past.
6. Menstrual disturbances/anorexia nervosa: present/past.
7. Get one week's baseline of food eaten, where and when.

Treatment possibilities
1. Self-monitoring, i.e. write down what they will eat at beginning of day and stick to it (include calorie amount if over-weight).
2. Stimulus-control programme to modify meal quantity, frequency and type of food eaten.
3. Cue exposure and response prevention to favourite foods.
4. Covert sensitization.

Measures

All usual measures and a graph of daily amount of food eaten during
and between meals and of vomiting, e.g.

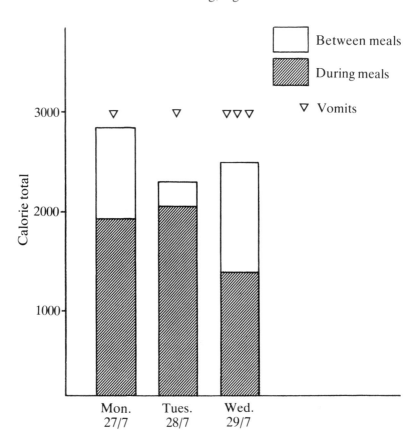

3.3.4 Anger Management

Assessment
1. Specific behavioural cues—current, predictable, e.g. being
   criticized, crowded situations.
2. Associated anxiety, anger, etc. and resultant avoidance.
3. Behaviour—excessive: shouting, swearing, physical aggression
                 deficit: listening, requests, keeping to topic.
4. Resultant feelings/consequences, i.e. isolation, rejection, guilt, etc.
5. Possible alternative actions based on deficit behaviour.

6. Police record for violence.
7. Threat of divorce, removal of children, etc.

Measures
Based on: *a*. Deficit behaviour.
        b. Avoidance.

3.3.5 Habit Disorders

Assessment
1. Duration of problem, detailed description of act.
2. Frequency of urges/acts (patient + third-party ratings).
3. Fluctuations, precursors, modifiers, thoughts while performing act.
4. Consequences of act—aversive, pleasant.
5. Reaction of relatives/friends.
6. Baseline measurement of frequency (e.g. dysfluencies per minute, tics per 5 minutes).

3.3.6 Sexual Problems

i. Assessment: heterosexual
Coital frequency, behaviour, fantasies and orgasm, now/past.
First coitus and response.
Engagements/marriages/separation.
Length and number of other relationships and accompanying sexual activity.
Masturbation frequency and fantasies now/past.
First masturbation/wet dream/menarche/petting.
Unpleasant sexual experiences.
Parents' attitude to sex.
Deviant behaviour.

ii. Assessment: deviant
Type(s) of deviant practice(s) and with whom? homosexual/paedophile/exhibitionist/transvestite/transsexual/sadist/masochist/voyeur/frotteur/fetish.

Frequency of deviance:	as urges	now/past
	as masturbation fantasies	now/past
	carried out in real life	now/past

Deviant experiences mainly with: regular partner/regular + casual/casual/solo.

Pattern of present deviant behaviour and modifying factors.
First deviant urges and practices.
Change in deviant practices over time.
Are family/friends aware of deviance?
Preferred reaction from other person(s) if any?
Feelings following act, e.g. remorse, disgust, relief, etc.
Self-control over urges, frequency, how done.
Arrests, convictions, probation orders, prison sentences, any charges pending?
Worst possible way and consequences of being caught.
Reason(s) for requesting change in behaviour.
Discharge: for premature ejaculators, report pre- and post-time to ejaculation in coitus.

iii. Additional measure for covert sensitization
1. Arousal from deviant imagery
2. Aversion from covert sensitizing imagery

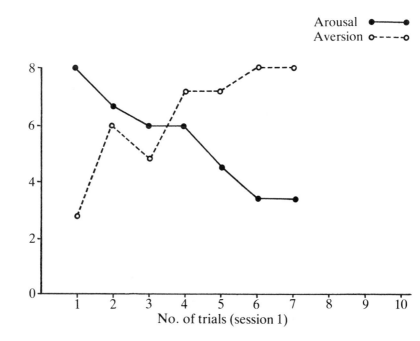

3.3.7 Sexual Dysfunction—Management Outline

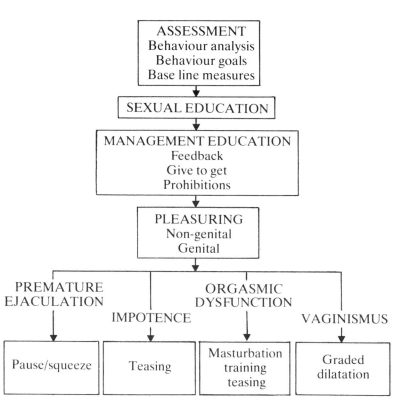

3.3.8 Social Skills Groups—Management Hints

Selection
—specific social anxieties
—specific social skills defects
—not depressed
—not paranoid
—not drugged or drunk

Preparation
—each patient individually
—each therapist (a good model?)

Session arrangements
—long and frequent early in treatment (e.g. 2 hours alternate days)
—limited total (e.g. 10) number of sessions
—rules, rewards, and sanctions
—maximum of 6 patients per group

Skill teaching
—demonstrate often
—small components
—positive feedback
—repetition
—audiovisual aids

Generalization
—outings in session
—immediate homework
—immediate reports (telephoned)
—relatives and friends
—contract cards

Evaluation
—concrete
—cumulative

Ending
—continue homework for up to 2 years
—half-way groups
—self-help
—follow-up

SPECIMEN LETTERS

3.4.1 Specimen Letter: Spider Phobia plus Rituals—Assessment

27.7.83

Dear Dr ....

Re. .... d.o.b. 3.3.33 Hosp. No.......

Address . . . .

Further to our letter of 19.7.83, this patient was seen yesterday for detailed assessment. Her main problem is fear of spiders of 45 years' duration which leads to extensive cleaning and checking rituals. She sought treatment recently because of a car accident she had a year ago after she suddenly discovered a spider in her car. Once she sees a spider she is unable to leave it in the home and becomes agitated with palpitations, sweating, a lump in the throat and frequency of micturition. She has developed an elaborate way of hoovering them up and sealing them inside with tissues inserted in the nozzle. Up to 65% of her daily activity is involved in checking for the presence of spiders. If she spots a spider she involves the whole family in the rituals of removing it, spraying pesticides and taping up cracks in the floor. She frequently seeks reassurance from her husband and asks him to check every nook and cranny of her home for spiders. As you mentioned in your letter, she could harm herself if startled in a situation requiring concentration. She is on no drugs, and does not abuse alcohol.

She lives in a semi-detached cottage in the country with her husband and one of her sons. She has good links with her neighbours and enjoys an active social life.

Her mother lives in Malaga and is aged 81 and well; she has always been a housewife and is 'uncommunicative but nice enough'. Her father died aged 64 in 1962 of renal failure; he was an exporter of coffee and a 'strict Victorian'. She is the third of eight siblings and they are all well. There is no family history of psychiatric disturbance. Her childhood was enjoyable; she left school at age 15 and went to college in Malaga, where she obtained a B.A. in Commerce. She is happily married, sexual adjustment is normal, and she has two sons. The family is sympathetic to her problem and very supportive.

At interview she was communicative and co-operative. She was unhappy about her problem, but not depressed and there was no evidence of psychosis. She understands the treatment plan and is keen to start.

Management of her spider phobia will be on an outpatient basis. It will be treated by graded prolonged exposure to spiders, plus a ban on cleaning and checking rituals and on seeking of reassurance. Homework will be set between sessions and her son will assist as co-therapist and liaise with her husband at all stages.

We will write again at completion of treatment.

Yours sincerely,

Nurse Therapist   Senior Registrar
Professor Marks' Unit

Specimen Letter: Spider Phobia plus Rituals—Discharge

31.10.83

Dear Dr . . . .

Re. . . . . d.o.b. 3.3.33 Hosp. No. . . . .

Address . . . .

Further to our letter of 27.7.83, this patient has been discharged from treatment. The main problem treated was a fear of spiders of 45 years' duration that led to extensive cleaning and checking rituals. She was treated with 6 sessions of graded prolonged exposure to real spiders. Her son acted as a co-therapist.

The spider phobia and rituals are much improved. She can now clean the house without calling for assistance, pick up spiders and place them outside, and look at TV programmes and books and magazines involving spiders. She sleeps better without nightmares of spiders and no longer uses any pesticides around the house. Her cleaning rituals have stopped and she no longer checks for spiders. Her friends and relatives have commented on the improvement and she recently returned to Malaga and impressed her family by her tolerance of spiders there. We have advised continued practice of exposure to spiders.

Recently she has had troublesome menopausal flushes and dizziness, but feels that she will cope with them.

We will write again after seeing her for 1-month follow-up.

Yours sincerely,

Nurse Therapist
Professor Marks' Unit

Specimen Letter: Spider Phobia plus Rituals—1-month Follow-up

13.12.83

Dear Dr . . . .

Re. . . . . d.o.b. 3.3.33 Hosp. No. . . . .

Address . . . .

Further to my letter of 31.10.83, this patient was seen for 1-month follow-up. She contines to maintain her gains and can handle spiders without difficulty. The cleaning and checking rituals also remain minimal so she has extra free time for leisure pursuits.

We will see her again for 3-month follow-up and will write to you then.

Yours sincerely,

Nurse Therapist
Professor Marks' Unit

Specimen Letter: Spider Phobia plus Rituals—3-month Follow-up

13.2.84

Dear Dr . . . .

Re. . . . . d.o.b. 3.3.33 Hosp. No. . . . .

Address . . . .

Further to my letter of 13.12.83, this patient was seen for 3-month follow-up. Her main problem of spider phobia remains much improved and she no longer checks for their presence or excessively cleans the house. There are no handicaps from this problem.

She was cheerful at interview despite severe knee pain which is being attended to at Guy's Hospital. She is determined to continue practising her exposure exercises by confronting spiders.

She will be seen again for 6-month follow-up and we will write to you again then.

Yours sincerely,

Nurse Therapist
Professor Marks' Unit

3.4.2 Specimen Letter: Phobias of Being Alone with Ectopic Heart Beats—Assessment

7.2.85

Dear Dr John,

Re. Mrs 'A'   d.o.b. 19.3.40   Hosp. No. . . . .

Address . . . .

Thank you for referring this patient, who was seen today for assessment for possible behavioural psychotherapy. Her main problem, of 16 years' duration, is anxiety and panic associated with bouts of missed (?ectopic) heart beats. These have produced fears of cardiac abnormality and led her to avoid being alone at any time. She constantly fears the onset of a 'bout' of heart beats. These are infrequent and very unpredictable, the last two occasions being 9th January and 31st December. The longest continuous period was in 1975 for 1 week.

An attack of missed heart beats starts with a sensation that her heart has stopped, followed by a sharp bang and then a fluttery feeling in her chest and throat, and two or three normal heart beats before the cycle repeats. When this occurs she becomes breathless, shaky, cries, shouts and may throw and break objects in her room. Such activity lasts between 15–45 minutes, after which she takes 5 mg diazepam and tries to sleep. She dreads a recurrence of further bouts and is terrified of dying through cardiac arrest.

Onset occurred in 1968; after being stung by a wasp she collapsed, was taken to hospital by ambulance, and was given intravenous adrenaline. As she recovered consciousness she became aware of her heart beating rapidly; at the time she was staring at a cardiac resuscitation trolley and heard someone say 'Oh . . . she's dead!' She was unable to reply and the whole experience was extremely frightening.

She is currently on oxprenolol 40 mg t.d.s., disopyramide 100 mg t.d.s. and diazepam 5 mg t.d.s. We advised her to discontinue the oxprenolol and diso-pyramide, as these might interfere with treatment.

Past treatments include admission to Barnet General Hospital for observation in 1973 for three days, for acute chest pain later diagnosed as stress. Assessment by Dr 'B' at Addenbrooke's Hospital in June 1984 revealed no cardiac abnormality. She has taken diazepam 10 mg/d almost continuously since 1975, recently reduced to 5 mg/d and takes an extra 5 mg 2–3 times a week when anxious, plus oxprenolol 40 mg since July 1984. Since December 1984 she has also had disopyramide, which seems to have reduced the attacks.

Her husband and son are aware of her problem but find it hard to understand. She lives with them in their own cottage. Her mother is aged 80 and in good health. They are in regular contact but her mother is irritable and difficult to deal with. Her father died in 1943 and she remembers little about him. She was born in Malaysia and moved to England when her father died.

Her childhood was difficult as she was always the centre of mother's attention. She left school at age 16 with no exams and worked as a hairdresser until she expected her first child. She married at age 20 and the marriage is happy. Her husband, aged 58, retired three years ago after selling his road-haulage business. He is in good health. They have a daughter aged 22 who is at university and son aged 19 awaiting entry to the Marines, both in good health.

Management of her fears of cardiac abnormality will be as an outpatient, involving exposure to her fears. She will be asked to 'bring on' her attack with accompanying anxiety and then to practise coping tactics to tolerate this (e.g. breathing exercises). She will also be asked to carry out exposure to routine daily tasks alone, e.g. shopping in a busy supermarket and exercise; exposure homework would be carried out between sessions, to consolidate treatment gains. We hope to involve her husband as a co-therapist.

We will write again at discharge.

Yours sincerely,

Nurse Therapist   Senior Registrar
Psychological Treatment Unit

Specimen Letter: Phobias of Being Alone with Ectopic Heart Beats—Discharge

13.6.85

Dear Dr John,

Re. Mrs 'A'   d.o.b. 19.3.40   Hosp. No. . . . .

Address . . . .

Further to our letter of 7.2.85, this patient has been discharged from treatment. Her main problem had been a fear of cardiac abnormality associated with bouts of ectopic beats, which led her to avoid being alone at any time, discontinue all activity during bouts of ectopic beats and seek constant reassurance from her family.

Treatment consisted of four sessions of exposure in imagination to the feared situation and one session where she viewed a film on the effects of long-term use of benzodiazepines. She was taught anxiety-coping tactics and exercised daily as 'homework' practice, especially during bouts of ectopic beats. She was also asked to gradually reduce all medication.

The problem has improved greatly and, although she still has bouts of ectopic beats, she can now continue any activity during them, regarding the ectopic beats as a minor discomfort. She regularly spends time alone and enjoys a very active social life. Her husband acted as co-therapist; he confirms that she is improved and no longer seeks reassurance from the family. She has discontinued all medication except diazepam 5 mg *mane*, which she has agreed to reduce and finally discontinue over the next few weeks.

She was advised to continue her exercise programme and coping tactics, especially during bouts of ectopic beats, to consolidate treatment gains.

We will write again after seeing her for 1-month follow-up.

Yours sincerely,

Nurse Therapist
Psychological Treatment Unit

**Specimen Letter: Phobias of Being Alone with Ectopic Heart Beats—
1-month Follow-up**

22.7.85

Dear Dr John,

Re. Mrs 'A'   d.o.b. 19.3.40   Hosp. No. . . . .

Address . . . .

Further to our letter of 13.6.85, this patient was seen again for 1-month
follow-up. Her main problem had been her anxiety associated with bouts of
ectopic beats, inability to be left alone at any time, and discontinuation of any
activity during bouts of ectopic beats.

She continues to be able to be by herself and to continue with activities during
bouts of ectopic beats, but complains of increased frequency and lengths of
time of these bouts with increased anxiety.

We tried to record these ectopics on an ECG but found it to be normal. Since
then we received a copy of the letter Dr 'A' wrote to you noting that he
recorded ventricular ectopics with no evidence of cardiac failure on 5.5.84. He
suggested that she discontinue smoking and taking alcohol. She has stopped
drinking alcohol but continues to smoke 20 cigarettes a day and is reluctant to
discontinue these.

We advised her to continue with the exercise programme and will encourage
her to stop smoking. She continues to take diazepam 5 mg *mane*.

We will write again after 3-month follow-up.

Yours sincerely,

Nurse Therapist
Psychological Treatment Unit

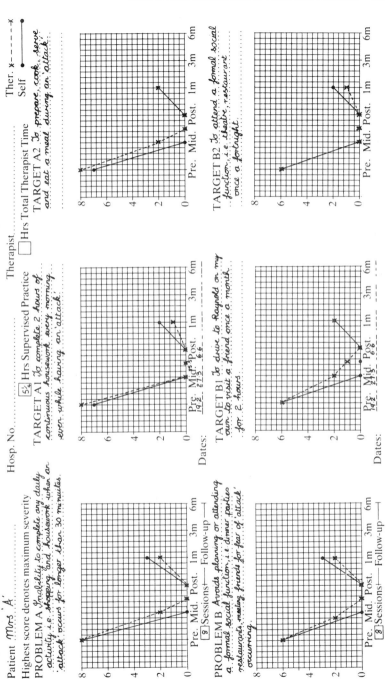

Patient *Mrs. 'A'*

Hosp. No............. Therapist............. Ther. ×– – – –×

Highest score denotes maximum severity □ Hrs Total Therapist Time Self ●————●

PROBLEM A *Inability to complete any daily activity i.e. shopping and housework when an attack occurs for longer than 30 minutes.*

TARGET A1 *To complete 2 hours of continuous housework every morning even while having an attack.* 5½ Hrs Supervised Practice

TARGET A2 *To prepare, cook, serve and eat a meal during an attack.*

PROBLEM B *Avoids planning or attending a formal social function, i.e. dinner parties, restaurants, seeing friends for fear of 'attack' occurring.*

TARGET B1 *To drive to Reynold on my own, to visit a friend once a month for 2 hours.*

TARGET B2 *To attend a formal social function, i.e. theatre, restaurant once a fortnight.*

Patient . . *Mrs 'A'* . . . .   Hosp. No. . . . .   Therapist . . . .   SELF RATING
Highest score denotes maximum severity

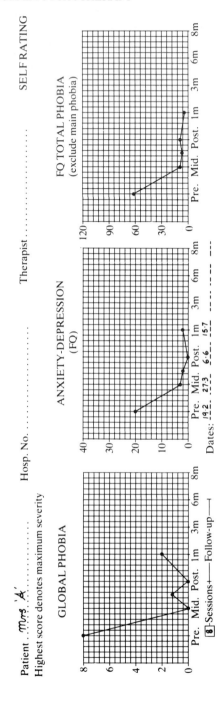

GLOBAL PHOBIA

ANXIETY-DEPRESSION
(FQ)

FQ TOTAL PHOBIA
(exclude main phobia)

Pre. Mid. Post. 1m   3m   6m   8m

Pre. Mid. Post. 1m   3m   6m   8m
Dates: 19.2 27.3 6.6 15.7 – – – – – – – – – – – – – –

Pre. Mid. Post. 1m   3m   6m   8m

[8]-Sessions———Follow-up———|

Patient . . . . Mrs 'A' . . . .   Hosp. No. . . . .

Highest score denotes maximum severity

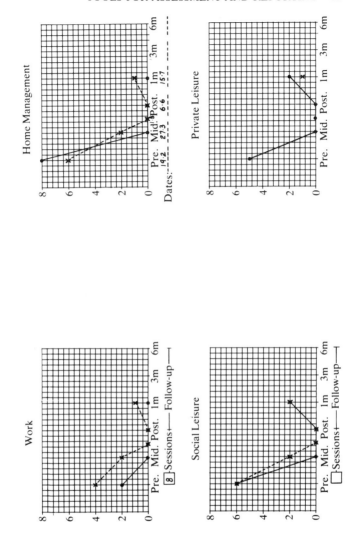

3.4.3 Specimen Letter: Injection Phobia—Assessment

5.12.84

Dear Dr 'C',

Re. Mrs 'B'   d.o.b. 10.12.49   Hosp. No. . . . .

Address . . . .

This patient was seen yesterday for assessment. Her main problem is fear and avoidance of injections, of lifelong duration. Since 1982 she has required twice-daily injections of insulin for her diabetes and her fear of injections has become severe. She prevented her children, now aged 15 and 14, from having any routine inoculations. She avoided injections of insulin on three occasions and was then admitted to hospital. Currently she has an abdominal catheter in situ; she slightly prefers this to subcutaneous injections. Two District Nurses visit her twice daily to administer insulin. Since 1982 she has been depressed about this problem and felt that she could not face a life involving daily injections. As a result, she has three times taken a drug overdose (February 1983, October 1983 and January 1984). She also tried to jump from Tower Bridge and London Bridge in October 1984. She does not abuse alcohol, and is on no medication apart from insulin.

She was treated for postnatal depression, in Warlingham Park Hospital in 1970 and again in 1972. She was again an inpatient there for depression in 1983, and an outpatient until April 1984. These admissions were related to her fear of injections and her diabetes.

She lives in a privately owned two-bedroomed house with her husband. Their son of 14, stays with them on alternate weekends. He attends a boarding school because 'he is hyperactive'. Their daughter of 15 attends a Save the Children school, and resides at a Children's Home. She feels guilty about her children living away from home and worries that her son is also becoming phobic of injections. She is unhappy in her marriage as her husband does not seem to understand her problems or give adequate support. She does not feel that the marriage is in danger of breaking up. She is involved in church activities, principally the junior choir, but has few other social links.

Her foster parents live in Hastings. Her relationship with them is fair. Her foster father is a retired ceramics worker in poor health. Her foster mother fostered 97 children in 37 years. The patient visits them 4–5 times a year. Her natural mother abandoned her at an early age, a traumatic experience; since then the patient has only once contacted her natural family. She discovered a brother in May 1984 and phones him twice weekly.

Her childhood was sheltered and strict. Her development was normal, although she had few school friends. Sexual development was normal. She has dysmenorrhoea and is presently receiving treatment for this.

At interview the patient was communicative and co-operative. She looked anxious, grimaced at the mention of injections, and said she is depressed. Her past psychiatric history is a cause of concern. She understands the treatment plan and is keen to start.

Management of her injection phobia will be on an outpatient basis by graded, prolonged exposure to syringes, needles and, ultimately, injections. Exposure homework involving needles etc., will be set between sessions.

We will write again on discharge.

Yours sincerely,

Nurse Therapist   Senior Registrar
Psychological Treatment Unit

Specimen Letter: Injection Phobia—Discharge

14.5.85

Dear Dr 'C',

Re. Mrs 'B'   d.o.b. 10.12.49   Hosp. No. . . . .

Address . . . .

Further to our letter of 5.12.84, this diabetic lady has been discharged from treatment. The main problem treated was her fear and avoidance of injections, of lifelong duration and resulting in admission to hospital for insulin infusion on three occasions. She was treated by 11 sessions of graded exposure to syringes, injections and venepuncture.

The injection phobia is much improved. She now self-injects insulin twice daily with only minimal anxiety. She is sleeping better without nightmares. Her daughter has commented on her improvement. We advised continued exposure to injections and syringes.

We will write again after seeing her for 1-month follow-up.

Yours sincerely,

Nurse Therapist
Psychological Treatment Unit

Specimen Letter: Injection Phobia—1-month Follow-up

19.6.85

Dear Dr 'C',

Re. Mrs 'B'   d.o.b. 10.12.49   Hosp. No........

Address ....

Further to our letter of 14.5.85, the above was seen for 1-month follow-up. She continues to maintain her gains and injects herself with insulin twice daily, without difficulty. She attended her dentist several times with only minimal anxiety.

We will see her again for 3-month follow-up, after which we will write to you again.

Yours sincerely,

Nurse Therapist
Psychological Treatment Unit

Specimen Letter: Injection Phobia—3-month Follow-up

4.9.85

Dear Dr 'C',

Re. Mrs 'B'   d.o.b. 10.12.49   Hosp. No........

Address ....

Further to our letter of 19.6.85, the above was seen for 3-month follow-up. Her injection phobia has returned and she has not self administered insulin for 4–5 weeks. She suffered an inflammatory/dermatitis? condition on her hands and was unable to give injections, following this she was unable to do so due to high anxiety. However, she has continued to attend her dentist with only minimal anxiety.

Her injection phobia will be treated with 2–3 sessions of prolonged exposure to subcutaneous injection and venepuncture. Homework including needles and self administration of insulin injections will be set between sessions.

We will write again.

Yours sincerely,

Nurse Therapist
Psychological Treatment Unit
cc G.P.—Dr 'D'

Specimen Letter: Injection Phobia—8-month Follow-up after Booster Treatment

14.1.86

Dear Dr 'C',

Re. Mrs 'B'   d.o.b. 10.12.49   Hosp. No. . . . .

Address . . . .

Further to our letter of 4.9.85, the above was seen for 8-month follow-up. Her injection phobia has much improved. She received 7 'booster' sessions of exposure to injections. She now administers all her insulin injections, despite difficulties from sharply fluctuating blood sugar levels. She has also discussed with the district nurses how to reduce their input.

However, she remains very anxious about venepuncture, and is unwilling to carry out all her exposure exercises for this. We have stressed the importance of continued exposure to injections and puncture by needles to maintain treatment gains.

We will see her in 1 month's time and will write again.

Yours sincerely,

Nurse Therapist
Psychological Treatment Unit

Specimen Letter: Injection Phobia—9-month Follow up

Dear

Re. Mrs 'B'   d.o.b. 10.12.49   Hosp. No. . . . .

Address . . . .

Further to our letter of 14.1.86, the above was seen for 9-month follow-up. Her injection phobia remains improved. She injects herself twice daily with insulin injection with only occasional anxiety. She is to make arrangements with the district nurses with a view to reducing this input, with an eventual discontinuation of their input. We have stressed the importance of continued exposure to needles and skin puncture.

We will not be seeing her again.

Yours sincerely,

Nurse Therapist
Psychological Treatment Unit

Patient ....

Highest score denotes maximum severity

PROBLEM A *Fear of injections leading to avoidances, resulting in health risk.*

Hosp. No. ....

Therapist ....

☐ Hrs Supervised Practice

TARGET A1 *To inject insulin twice daily without undue anxiety.*

Ther. ✗– – – –✗ Self ●————●

☐ Hrs Total Therapist Time

TARGET A2 *To be able to have 5 ml. of blood withdrawn from a vein.*

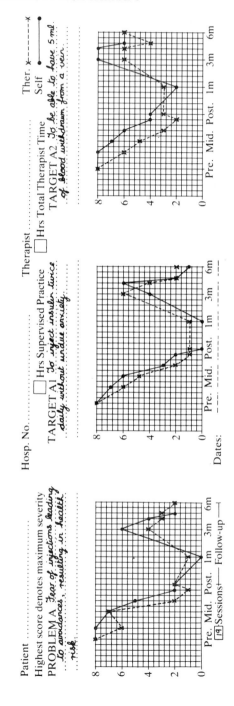

Dates:

◻ Sessions † ——— Follow-up ———

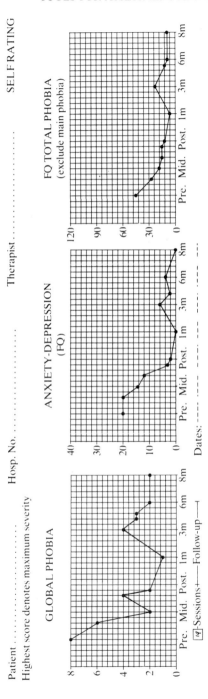

Patient . . . .

Highest score denotes maximum severity

Therapist . . . .

Hosp. No. . . . .

Self   ●——————●

Ther.   ✕- - - - - - - -✕

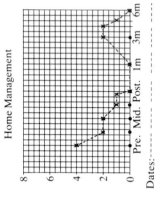

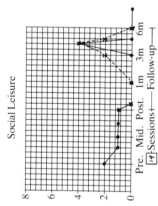

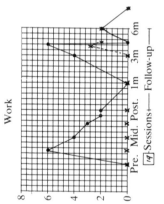

3.4.4 Specimen Letter: Obsessive-compulsive Disorder—Assessment

15.3.85

Dear Dr ....

Re. Mr 'D'   d.o.b. 13.6.50   Hosp. No: ....

Address ....

Further to my letter of 18.2.83, this patient was seen again yesterday for detailed assessment. His main problem is fear of contamination by dirt and dog excreta, and of contaminating his family with herpes, following a herpetic eye infection 15 months ago. He fears contracting or transmitting disease, particularly blindness, though he knows this is irrational. These fears began 1 year ago leading to repeated prolonged washing rituals which increased over the last 6 months. The patient avoids situations where excreta may be present such as local parks and his garden, and avoids touching or using muddied shoes. Touching contaminated articles leads to repeated washing of his hands 25 to 30 times a day for fear of contaminating others. He spends at least 1½ hours bathing daily, washes his hair daily, soaping and rinsing 5 times, and ensures that his clothing is kept separate from that of his wife and children, washing all such clothing if he accidentally touches it. These activities, along with seeking reassurance from his wife, occupy about 5 hours each day. There are also checking rituals concerning locks, gas, electricity, taps and household tasks, with accompanying anxiety but no obsessive thoughts. These also date from 1 year ago.

He has never been a psychiatric inpatient, but was treated in outpatients at this hospital in 1967, 1971 and 1982. Each time the precipitating factor was fear of illness and he was said to have depression within the context of an obsessive personality. Present medication is amitriptyline 150 mg *nocte*, which he is keen to discontinue. There is no significant medical history other than childhood asthma, though he has frequently sought reassurance from various medical agencies for minor complaints.

Mr 'D' now lives with his wife and 2 children in a two-bedroomed house which he owns. His father died 5 years ago and his mother is alive and well. He is an only child. There is no family history of mental illness. He was a spoiled child and overprotected because of his asthma. He attended school until 15, disliking it because he felt a failure after not gaining a grammar school place, and left without qualifications. Since then he has passed 4 'O' levels by private study. He is employed as a postal worker, a job he dislikes, but is not considering changing it.

Puberty was at 12 and first coitus at 13. The patient had several long relationships before marrying at 24 and the marriage is happy, with satisfying sex.

At interview, Mr 'D' was neatly dressed, scrupulously clean man who gave information readily, if in excessive detail. Mood was low. Appetite, sleep, libido and interest were normal. There was no evidence of psychosis.

Treatment of the compulsive rituals will be as an outpatient, involving graded exposure to contamination coupled with a ban on rituals. He understood the rationale and treatment and was eager to begin, despite apprehension at the prospect. His wife was also seen and will act as co-therapist.

We will write again at discharge.

Yours sincerely,

Nurse Therapist   Senior Registrar
Professor Marks' Unit

Specimen Letter: Obsessive-compulsive Disorder—Discharge

5.5.83

Dear Dr . . . .

Re. Mr 'D'   d.o.b. 13.6.50   Hosp. No. . . . .

Address . . . .

Further to our letter of 15.3.83, this patient has now been discharged from treatment. The main problems treated were excessive washing, bathing and cleaning due to fear of contamination by dirt and excreta, and excessive checking. His dissatisfaction with his present work and social life was not treated directly.

He had four sessions of exposure to 'contamination', plus a ban on rituals. His wife was involved as co-therapist and he carried out self-exposure tasks at home under her supervision.

The compulsive rituals have greatly reduced. He used to spend up to 5 hours a day washing and bathing; this is now less than 45 minutes and he no longer checks more than once before going to bed or asks his wife for reassurance. He is now able to take his children to the park and do gardening. He has more free time since he no longer repeats activities.

Residual problems are minor anxiety about dirt and illness that no longer lead to rituals, and continuing dissatisfaction with his work and social life. He has been advised to continue confronting his feared situations and has been given behavioural advice regarding expansion of his social network.

We will write again after seeing him for 1-month follow-up.

Yours sincerely,

Nurse Therapist
Professor Marks' Unit

Specimen Letter: Obsessive-compulsive Disorder—Follow-up

5.6.83

Dear Dr ....

Re. Mr 'D'   d.o.b. 13.6.50   Hosp. No........

Address ....

Further to our letter of 5.5.83, this patient has now been seen for 1-month follow-up. His anxieties regarding dirt and illness are further reduced and he has made successful efforts to expand his social network. The washing and checking rituals are now absent. He feels able to see things in proportion and is pleased with his progress.

We will write again after 3-month follow-up.

Yours sincerely,

Nurse Therapist
Professor Marks' Unit

3.4.5 Specimen Letter: Social Skills Deficit—Assessment

2.12.83

Dear Dr ....

Re. Miss 'E'   d.o.b. 1.2.57   Hosp. No........

Address ....

Thank you for referring this patient, who was assessed today for possible behavioural psychotherapy. Her main problem, of 8-years' duration, is social anxiety and inability to talk or disagree with others, particularly women of her own age. At work she is uncommunicative and slow and fears she may lose her job because of the problem. She improved following group social skills training at Guy's Hospital in 1978, but in 1980 became acutely anxious and was admitted to St Luke's, Woodside for 5 months, where she was treated by drugs and group psychotherapy. She again improved, but has relapsed once more. She currently takes no medication or excess alcohol.

Miss 'E' lives with her parents and two younger brothers in a semi-detached house. The home atmosphere is 'fraught' due to the elder brother's 'temper tantrums'; she is considering moving to a hostel as she did on a previous occasion to avoid family conflict. There is nothing of note in the medical or psychiatric history of the family.

She was an anxious child with few friends, due to her communication difficulties. She has worked in her present and only job since 1974, when she left school, but has never obtained promotion. She has had a few casual relationships with men, the longest lasting for 6 months, in 1980. He was her only sexual partner, the frequency of sexual intercourse being about twice a month and she was orgasmic on 50% of the occasions. She goes out socially once a week to an 18+ club, where she has a casual boyfriend.

At interview, she was initially very anxious, fidgeting constantly and often unable to answer questions or to complete sentences. Eye contact was extremely poor and speech was quiet, monotonous and retarded. She gradually relaxed and became more fluent in speech. There was no evidence of psychosis or depression.

Management of her social anxiety and skills deficits will be as an outpatient by means of group social skills training with adjunctive coping tactics for the anxiety. She will be asked to consolidate treatment gains by a series of homework tasks. She has no one who can act as a co-therapist.

We will write to you again at discharge.

Yours sincerely,

Nurse Therapist   Senior Registrar
Professor Marks' Unit

Specimen Letter: Social Skills Deficit—Discharge

26.1.84

Dear Dr ....

Re. Miss 'E'   d.o.b. 1.2.57   Hosp. No........

Address ....

Further to our letter of 2.12.83 this patient has now been discharged from treatment. The main problem treated was her social anxiety and inability to communicate with others, particularly colleagues at work. Treatment consisted of 2 individual sessions of practice in anxiety-coping tactics and 9 sessions of group social skills training. This involved role play of everyday social situations with feedback from peers and therapists. She was asked to practise the social skills taught during the sessions as 'homework' in real life situations between sessions.

She has gained in confidence from the group and now converses more naturally with her work colleagues. Family conflict has worsened and she continues to think of living away from the family.

She has been advised to continue practising the social skills she has acquired in real life situations, for though there is improvement, her eye contact remains poor and her anxiety is high at times.

We will write again after 1-month follow-up.

Yours sincerely,

Nurse Therapist
Professor Marks' Unit

Specimen Letter: Social Skills Deficit—1-month Follow-up

29.2.84

Dear Dr . . . .

Re. Miss 'E'   d.o.b. 1.2.57   Hosp. No. . . . .

Address . . . .

Further to our letter of 26.1.84, this young woman has now been seen for 1-month follow-up. The main problem treated was her anxiety and inability to communicate with others, particularly colleagues at work.

She has maintained all her treatment gains; she reports continued improvement at work and is now an active union member. She is still looking for a flat as the home conflict continues. Her eye contact has improved and she is less anxious when conversing. She has been advised to continue practising the social skills she learned in treatment and to increase her social opportunities.

We will write again after 3-month follow-up.

Yours sincerely,

Nurse Therapist
Professor Marks' Unit

Specimen Letter: Social Skills Deficit—3-month Follow-up

2.5.84

Dear Dr . . . .

Re. Miss 'E'   d.o.b. 1.2.57   Hosp. No. . . . .

Address . . . .

Further to our letter of 29.2.84, this young woman has been seen for 3-month follow-up. She has maintained all her treatment gains and reports continued improvement at work. She has now moved into a hostel and is increasing her social contacts.

We will write to you again after 6-month follow-up.

Yours sincerely,

Nurse Therapist
Professor Marks' Unit

3.4.6 Specimen Letter: Sexual Dysfunction—Assessment

17.11.83

Dear Dr ....

Re. Mr and Mrs 'E'   d.o.b. 26.6.57 & 2.2.62   Hosp. No........

Address ....

Further to my letter of 20.7.83 this couple were seen for detailed assessment on 9.11.83. The reason for this delay was that they wished to live in their own home before starting treatment. Their main problem is Mr 'E''s premature ejaculation on all coital occasions and Mrs 'E''s coital anorgasmia with similar frequency; both problems have been present since the start of their relationship five years ago. Currently, coitus takes place 4 times per week. Mr 'E' usually ejaculates about 30 seconds after penetration, though he can delay ejaculation for up to 3 minutes if his wife takes the female superior position. She is orgasmic on all occasions of mutual masturbation, which takes place 80% of the time after and 20% before coitus. They both enjoy foreplay prior to coitus but rush it a bit as they are anxious that he may ejaculate before penetration. They have oral sex once a week when both are orgasmic, masturbate mutually to orgasm 4 times per week, and also masturbate individually. Both are orgasmic on self-masturbation after which he can delay ejaculation for up to 5 minutes. They have tried the 'squeeze' technique in the past but could not carry it out properly.

Mr 'E' reached puberty at age 12 and masturbated since age 15. He had three casual girlfriends prior to his wife. First coitus was at age 20 with a casual girlfriend; he ejaculated prematurely. He had no formal sex education. His wife reached menarche at age 13. She initially had dysmenorrhoea but now has no menstrual problems. Sex education was given by her sister and friends. She had a few casual boyfriends before her husband and first coitus was at age 16, with him. They have been married for 15 months; first coitus took place four months into the relationship.

They have no children and live in their own 3-bedroomed house in Harlow. Joint activities are shopping and socialising with friends, and their relationship is close and happy.

Neither of them abuse alcohol or are on any medication. There is no past psychiatric history and their physical health is good.

Both had a happy childhood and get on well with their families. There is no family psychiatric history. Schooling for both was from age 5–16 and CSEs were achieved. Mr 'E' is employed as a shopfitting estimator and Mrs 'E' works for a bank.

At interview, they were smartly dressed, co-operative and informative. There was no evidence of depression or psychosis. Their orientation, concentration and memory were good. They related well together.

Management of their premature ejaculation and coital anorgasmia will be on an outpatient basis. Behavioural treatment will consist of sexual skills training, incorporating sensate focus and an initial ban on intercourse which will then gradually be reintroduced once they have learned to use the 'squeeze' technique.

We will write again at completion of treatment.

Yours sincerely,

Nurse Therapist   Senior Registrar
Professor Marks' Unit

Specimen Letter: Sexual Dysfunction—Discharge

                                                                    12.5.84

Dear Dr ....

Re. Mr and Mrs 'E'   d.o.b. 26.6.57 & 2.2.62   Hosp. No........

Address ....

Further to my letter of 17.11.83 this couple have now been discharged from treatment. I am writing to you, as they have not acquired a new G.P. and their records presumably remain with your practice. The main problem was Mr 'E''s inability to control ejaculation on all occasions of sexual intercourse, and his wife's coital anorgasmia with similar frequency. They had four sessions of instruction in sexual skills training including the 'squeeze' technique. Sexual homework was set between sessions and regular telephone contact was made to assess their progress at each stage.

The problem has improved greatly. Coitus takes place four times a week. On 75% of these occasions Mr 'E' is now able to control ejaculation for 5–7 minutes after penetration, his wife applying the 'squeeze' technique. On the remaining 25% he controls ejaculation for 2 minutes. Before treatment he ejaculated 30 seconds after penetration on all occasions. His wife says she is orgasmic during coitus on 50% of all occasions; pre-treatment she was orgasmic only during mutual masturbation.

The couple are very pleased with their progress and feel that they could eventually fade out the 'squeeze' technique. They know that on occasion premature ejaculation may recur and that if this happens they should resume practice of the 'squeeze' method. If that fails repeatedly they should resume practice with mutual masturbation until they feel ready to resume intercourse.

We will see them for 1-month follow-up and write to you again.

Yours sincerely,

Nurse Therapist
Professor Marks' Unit

Specimen Letter: Sexual Dysfunction—1-month Follow-up

20.6.84

Dear Dr . . . .

Re. Mr and Mrs 'E'   d.o.b. 26.6.57 & 2.2.62   Hosp. No. . . . .

Address . . . .

Further to my letter of 12.5.84 this couple were seen yesterday for 1-month follow-up. They continue to extend the gains made during treatment. They no longer need to use the 'squeeze' technique as he can now control his ejaculation on all coital occasions. Mrs 'E' is now usually orgasmic during coitus.

The couple are happy with their marital and sexual relationship. They think that they could use the sexual skills learned in treatment to overcome the problem should it recur in future.

We will see them again for 3-month follow-up and will inform you of their progress.

Yours sincerely,

Nurse Therapist
Professor Marks' Unit

3.4.7 Specimen Letter: Sexual Deviation—Assessment

4.6.83

Dear Dr 'Z' . . . .

Re. Mrs 'H'   d.o.b. 9.9.46   Hosp. No. . . . .

Address . . . .

Further to our letter of 12.5.83 this patient has been assessed and started treatment. Her main problem, of 30 years' duration, is of defaecating and urinating in her underwear while masturbating to orgasm. This problem has worsened over the last five years and there are no obvious stimuli for the activity. She does it on her own, putting on an old pair of pants pulled up tight, defaecating and urinating in them and rubbing her legs together without hand-to-genital contact. She gets most satisfaction from the soiling and recently had urges to do this in public but never during coitus. Currently she has urges to carry out the act 4 times a day and acts on them 3 times a week. The activity usually takes 10 minutes but can last up to half an hour. It began at age 6 after she soiled and stimulated outside the school lavatories to keep warm.

A second problem is that she has always found intercourse unstimulating and been anorgasmic. She has coitus 3 times a week and washes immediately after as she considers it 'messy'. She is not distressed by non-coital contact and loves her husband. Both problems cause her much guilt and anxiety. She also has sex for money with a friend of her husband who threatens to blackmail her if she stops, but she has decided to finish this nevertheless. In addition she has some lesbian feelings on which she has never acted and does not see these as a problem. Her husband, a garage mechanic, is five years older than her. She met him at age 17 and married a year later when 6 months pregnant. He is satisfied with their marital and sexual life. He learned about the sexual deviation recently and views it as a residual childhood disorder. Mrs 'H' is aware of strong sexual feelings towards her youngest child, a son aged 13, but has confided in her husband. Their three children are well.

Mrs 'H' lives in a 3-bedroomed semi-detached house with her husband and two of her three children. She works as a part-time home help and has good links with neighbours and voluntary organizations.

She has one older brother; two siblings were lost at birth. Her childhood was happy despite many arguments between her parents. Her father, aged 66, is a retired transport manager who suffers from rheumatism but is well. Mother, aged 65, is a housewife, 'quick-tempered', and has arthritis needing surgery. There is no family history of psychiatric illness. Her own childhood was unremarkable except for nocturnal enuresis until age 14.

At interview she was plump and neatly dressed and talked readily. Cognition was clear and although upset about her problem she was not depressed. There was no evidence of psychosis. She understood treatment and was keen to begin. Her husband was seen separately but refused to recount his own sexual history. His background and mental state were unremarkable. He is willing to co-operate in therapy, but regards the problem as being entirely his wife's.

Management of her sexual deviation will be on an outpatient basis. She will have covert sensitization, using an aversive image of her mother discovering her during deviant activity. For the anorgasmia she will have a programme of sexual education by talks, books and films; graded practice of sexual activity has begun, starting with non-genital sensate focus. She has been encouraged to take a lead in planning and monitoring both programmes and her husband will be a co-therapist at all stages.

As they live so far away from our unit the programme will continue for one month as sexual homework with the patient sending records to us weekly and telephoning regularly. We would appreciate your nomination of a local health worker (e.g. community nurse or psychologist) to contact us and be briefed on the techniques so that they can act as a co-therapist to allow continuity of care and follow-up after discharge.

We will write again in one month.

Yours sincerely,

Nurse Therapist   Senior Registrar
Professor Marks' Unit

Specimen Letter: Sexual Deviation—Re-referral

25.7.83

Dear Dr 'Y' . . . .

Re. Mrs 'H'   d.o.b. 9.9.46   Hosp. No. . . . .

Address . . . .

Thank you for your letter of 1.7.83 informing us that Mr 'J' of your Psychology Department is willing to complete the treatment programme for this patient.

The patient last attended this unit on 27.6.83 accompanied by her husband, and it was evident that the travel here was becoming a strain. She was pleased with her progress, had completed the sensate focus and is now having satisfactory coitus with her husband to orgasm 3 times a week.

Her deviant urges still occur 3 times a day but the frequency of deviant acts has dropped from 12 times to once a month. She was not depressed.

We phoned Mr 'J' and arranged to meet him here on 5.8.83. He will bring the patient with him for a full handover of her case.

Mrs 'H' has been advised to continue at home with her covert sensitization which she understands well. The prognosis is favourable if she continues her homework.

We will write again at discharge.

Yours sincerely,

Nurse Therapist
Professor Marks' Unit

Specimen Letter: Sexual Deviation—1-month Follow-up

25.7.83

Dear Dr 'Z' ....

Re. Mrs 'H'   d.o.b. 9.9.46   Hosp. No.......

Address ....

Further to our letter of 4.6.83 we have just received a 1-month report and follow-up ratings from this patient. She has improved in her main problem of urges and acts concerning defaecating and urinating in her underwear while masturbating. She feels better able to control these urges and coitus with husband is now more enjoyable. She is orgasmic on 75% of occasions and the frequency of coitus with her husband has increased from 3 to 4 times a week.

She is happy with her progress. We advised her to continue practice of covert sensitization to strengthen control of her deviant urges.

We will write again after 3-month follow-up.

Yours sincerely,

Nurse Therapist
Professor Marks' Unit

3.4.8 Specimen Letter: Eating Disorder—Assessment

17.5.83

Dear Dr ....

Re. .... d.o.b. 3.3.64 Hosp. No.......

Address ....

Further to our letter of 12.5.83 this patient was seen today for detailed assessment. Her main problem, of one year's duration, is a compulsion to eat about a pound of raw ground rice per day. She eats it straight from the bag in small mouthfuls, 20–30 at a time. No specific cues trigger her rice eating, although she eats more when unhappy or tense. She eats the rice alone as it embarrasses her. When in company she makes excuses to leave so that she can eat rice. She enjoys its taste and texture. When she cannot obtain rice she becomes anxious, sweats and has stomach pains and then substitutes something similar, e.g. breadcrumbs. She is happy with her present weight of 60 kg. Menses are regular. Although her rice eating can result in nausea and diarrhoea, she denies self-induced vomiting or abuse of purgatives. As you know, she is a vegetarian who likes few vegetables; this, together with the satiety produced by the rice, leads to a restricted diet. We contacted her dentist, who notes that her rice eating is damaging her teeth and gums.

The patient lives with her mother but their relationship has deteriorated since her parents' separation 18 months ago. Last year she was treated for an injection phobia by a nurse therapist from this unit. The phobia improved but she did not practise her exposure tasks to consolidate her gains and still has some residual symptoms.

She is an only child and remembers a spoilt but happy childhood despite having had asthma from an early age. She left school at 18 having attained 10 'O' and 2 'A' Levels, and is at present studying for a Diploma in technical illustration. At 17 she had post-abortion depression which improved spontaneously. She was anorgasmic during sexual intercourse with her boyfriend, whom she has known for 2 years. She declined treatment for this as she does not want him to know about it.

At interview she was casually dressed and anxious at first but had no difficulty in expressing herself. She was unhappy about her problem but not depressed and there was no evidence of psychosis. Memory and concentration were normal.

Management of the compulsive rice-eating will be as an outpatient. Initially, she has been asked to reduce her intake of rice from 1 lb to ¼ lb daily over the next 2 weeks and to chart her intake in a diary. She also wants to discuss the problems she is having with her parents. We agreed to these counselling sessions on condition that she reduces her rice eating and keeps up her diary.

We will write again in 2 weeks.

Yours sincerely,

Nurse Therapist   Senior Registrar
Professor Marks' Unit

Specimen Letter: Eating Disorder—Review

20.7.83

Dear Dr ....

Re. .... d.o.b. 3.3.64 Hosp. No. ....

Address ....

Further to our letter of 17.5.83 this patient was seen again for review of her excessive eating of raw ground rice. She has gradually reduced her rice intake from 1 lb to a ¼ lb daily and charted her progress in a diary. As agreed we commenced the counselling sessions which were contingent upon this self-regulation programme. Her diet is still rather restricted and she will see a dietitian for advice.

She insists on more supervised treatment to reduce the final quarter pound. We agreed to do this by covert sensitization; this involves her being trained to imagine an aversive fantasy when experiencing the urge to eat rice. The aversive fantasy is of continued rice eating leading to her becoming very fat and unattractive.

We will write again at discharge.

Yours sincerely,

Nurse Therapist
Professor Marks' Unit

Specimen Letter: Eating Disorder—Discharge

7.9.83

Dear Dr . . . .

Re. . . . . d.o.b. 3.3.64 Hosp. No. . . . .

Address . . . .

Further to our letter of 20.7.83 this patient has been discharged from active treatment. The main problem treated was a compulsion to eat about 1 lb of raw ground rice daily leading to neglect of her diet and damage to her teeth and gums.

The problem is now much improved. She has achieved the treatment target of not eating rice for 14 days. She is also eating a more balanced and varied daily diet and is halfway through a course of dental treatment. In addition she reports a better relationship with her parents. We have offered her the option of further supportive counselling and if she agrees we will refer her to an appropriate agency. We advised her to continue practice of her treatment exercises and will write again after 1-month follow-up.

Yours sincerely,

Nurse Therapist
Professor Marks' Unit

Specimen Letter: Eating Disorder—1-month Follow-up

5.10.83

Dear Dr . . . .

Re. . . . . d.o.b. 3.3.64 Hosp. No. . . . .

Address . . . .

Further to our letter of 7.9.83 this patient was seen today for 1-month follow-up. She has lost some of the gains made during treatment and only goes for

7 days without eating raw ground rice, after which she consumes about ½ lb over 3 days. She views this as acceptable and is happy with her progress. She still eats a varied and balanced diet, is receiving dental treatment and continues to have improved relations with her parents. She no longer wants further supportive counselling. I advised her to continue practice of the treatment exercises and will write to you again after 3-month follow-up.

Yours sincerely,

Nurse Therapist
Professor Marks' Unit

Specimen Letter: Eating Disorder—3-month Follow-up

10.1.84

Dear Dr ....

Re. .... d.o.b. 3.3.64 Hosp. No.......

Address ....

Further to our letter of 5.10.83 this patient was seen for 3-month follow-up. She has continued to lose the gains made in treatment and now eats about ½ lb of ground rice a day, which she feels is unacceptable.

She has been advised to continue practice of covert sensitization and we suggested a more potent aversive fantasy. We also suggested that she undertakes a competing response whenever she gets the urge to eat rice; in this case she is to peel and chew a carrot.

She continues to report improved relationships with her parents and has finished a course of dental treatment.

We will write again after 6-month follow-up.

Yours sincerely,

Nurse Therapist
Professor Marks' Unit

Specimen Letter: Eating Disorder—6-month Follow-up

10.4.84

Dear Dr ....

Re. .... d.o.b. 3.3.64 Hosp. No.......

Address . . . .

Further to our letter of 10.1.84 this patient was seen for 6-month follow-up. She has made good the gains lost at 3-month follow-up and has not eaten any ground rice for 2 months. Recently she moved into a flat with her boyfriend, who helped her a great deal with her problem. She has been encouraged to continue practising the treatment exercises and we will write again at 1-year follow-up.

Yours sincerely,

Nurse Therapist
Professor Marks' Unit

3.4.9 Specimen Letter: Stammer—Assessment

14.3.83

Dear Dr 'N' . . . .

Re. Mr 'P' . . . .   d.o.b. 29.7.40 Hosp. No. . . . .

Address . . . .

Further to our letter of 4.3.83 this patient was seen today for more detailed assessment. His main problem is stammering with at least 50% of all utterances. This dates to age 3 and he was teased at school. It is worse when using words beginning with b, p, m, s, w, u, l and v. He also stammers talking on the telephone and socially, e.g. asking for goods in shops and public houses. He is mildly anxious before talking and markedly so when stammering. He is an instrument technician and works off-shore on an oil rig, with alternate weeks on and off shore. His work requires regular conversation with other workers and his stammer frequently makes him ineffective and underachieving at work. He has been told that promotion prospects are poor.

He is reluctant to speak freely at work or socially. Past treatments include speech therapy at a clinic in Grimsby 1950–1952 using regulated breathing and alphabet charts; this was unsuccessful. In 1961 for 6 months he was seen by a consultant psychiatrist in Grimsby; he had drugs (details not known) and systematic desensitization with metronome pacing, with some success. Finally in 1975 for 6 months he had auditory voice feedback without improvement.

He lives in his own detached 4-bedroomed house with his wife and 4 children. She is happy with his work rotation. His father, age 74 and a director of an export company is 'strict and rigid'. His mother, aged 73 is 'a nice person'. He has a younger brother aged 39 to whom he is not close. There is no family history of mental illness. His early years were unremarkable apart from his stammer. He has been married for 21 years and sexual adjustment is normal.

At interview he was neatly dressed, tense and anxious. He spoke in a loud voice in a hurried manner with several speech dysfluencies. When asked to read for 3 minutes from a novel, 22 of 172 words were dysfluent. There was no evidence of depression or psychosis and memory and orientation were good. He was keen to have treatment.

Management of his stammering will be on an outpatient basis by means of regulated breathing exercises linked with practice in real life in those situations where most anxiety and dysfluencies occur. His wife and a friend working on board the oil rig are willing to become co-therapists. Speech-exercise home-work will be set between sessions. The initial treatment session at this unit will be 4 hours long. We will write again at discharge

Yours sincerely,

Nurse Therapist
Professor Marks' Unit

Specimen Letter: Stammer—Discharge

3.8.83

Dear Dr 'N' . . . .

Re. Mr 'P'   d.o.b. 29.7.40   Hosp. No. . . . .

Address . . . .

Further to our letter of 14.3.83 this patient has been discharged from treatment. His main problem was one of stammering in day-to-day situations. He had 5 sessions of treatment by regulated breathing linked to real life practice in stammering-prone and anxiety-evoking situations. Due to his shift system on board the oil rigs, telephone contact and homework were employed to ensure regular practice. His wife and a colleague on board the oil rig acted as co-therapists.

The stammering has improved and its frequency has been halved. He is now dysfluent with only 25% of utterances and this can be reduced further if he uses the breathing technique properly. His friends have commented on his success. He now answers phones readily and enters into conversation, both of which were previously avoided. He is generally less anxious and is pleased with his gains.

He feels that his residual dysfluencies result from his reluctance to use the regulated breathing techniques at work. Pressures on board the oil rig and the noise and social anxiety compound the problem. At home he is more relaxed and converses freely, getting praise and attention from all the family.

I have advised continued practice of the regulated breathing technique and speaking in the anxiety-evoking situations, to consolidate his gains. We will write to you again after 1-month follow-up.

Yours sincerely,

Nurse Therapist
Professor Marks' Unit

Specimen Letter: Stammer—1-month Follow-up

6.9.83

Dear Dr 'N' . . . .

Re. Mr 'P'   d.o.b. 29.7.40   Hosp. No. . . . .

Address . . . .

Further to our letter of 3.8.83 we have now managed to obtain follow-up information on this patient after delay due to his offshore duties.

He has continued to maintain his treatment gains. Using the regulated breathing approach he could read 408 words in 5 minutes with only 10 dysfluencies, representing a 5-fold improvement.

He no longer avoids social situations onshore and now goes into shops, asks directions, and orders drinks. Offshore, on the oil rigs, he still avoids certain phone calls due to 'the risk of misunderstandings'.

He does not practise the regulated breathing approach as consistently as he might but is pleased with his progress. He has been advised to continue practise in all areas and we will write to you again after 3-month follow-up.

Yours sincerely,

Nurse Therapist
Professor Marks' Unit

Specimen Letter: Stammer—3-month Follow-up

2.11.83

Dear Dr 'N' . . . .

Re. Mr 'P'   d.o.b. 29.7.40   Hosp. No. . . . .

Address . . . .

Further to our letter of 6.9.83 Mr . . . . recently wrote about his progress and returned completed 3-month follow-up forms.

There has been no additional progress, though he continues to practise regulated breathing. His condition fluctuates and in the last 2 months he had 3 weeks virtually free of dysfluencies. On average he can now read 420 words in 5 minutes with only 10–15 dysfluencies.

When onshore he continues to confront social situations without discomfort. His self-confidence has increased and he has stopped avoiding making phone calls.

His letter indicates that he is pleased with his progress and no longer feels handicapped. He assured us that he will continue to practise regulated breathing in social situations.

We will write again after 6-month follow-up.

Yours sincerely,

Nurse Therapist
Professor Marks' Unit

3.4.10 Specimen Letter: Irritable Bowel Syndrome—Assessment

26.7.83

Dear Dr 'L',

Re. Mrs 'M'   d.o.b. 2.11.56   Hosp. No. . . . .

Address . . . .

Thank you for your referral letter of 9.3.83 about this patient, who was seen today for possible behavioural psychotherapy. Her main problem is pain, distension and contractions of the bowel, infrequent and irregular bowel movements, of 18 months' duration. This is sometimes cued by anxiety in specific situations, e.g. during a rushed day at work or whilst shopping, and also by chronic tension and dissatisfaction. Pain is worst immediately prior to passing motions, which are small and well-formed. Desired frequency is daily. Sporadic aperients and a high-fibre diet had little effect. She currently takes Regulan granules to increase stool bulk, and Limbritol 10 (amitriptyline 25 + chlordiazepoxide 10), which has eased her general anxiety but not the bowel problem. Frequency of coitus has dropped from weekly to once every 6 weeks.

An unrelated problem is that she has never been orgasmic with her husband and this has led to tension between them. She is undecided whether to have this treated and a full assessment of it will be made should she choose to have it treated.

Mrs 'M' was initially treated by her GP for a 'pulled muscle', for 6 months, before being seen by Dr 'J', when irritable bowel syndrome was diagnosed. She was referred to yourself in October 1982 and thence, to Westminster Pastoral

Foundation for long-term psychotherapy, an option which she found unacceptable. There were two single consultations during her teens for mild depression, but there is no other past psychiatric history. Her medical history is uneventful.

The patient lives with her husband in their own 2-bedroomed terraced house. Their relationship is happy despite some tension from her problem, as she feels he does not understand. They have no children. Her own parents are alive and well and are separated. She rarely sees her father, following conflict from his marital breakdown. Her childhood and adolescence were marred by conflict at home. She is close to her mother. She is the youngest of 4 children, of whom 3 survive, and was always the happiest in the family. A sister suicided when the patient was 17 and a brother was hospitalized for depression at age 19.

Menarche was at age 12 and the patient had no sexual instruction. First coitus was at age 17. She has had 4 sexual partners and been with her husband for 6 years. They married 18 months ago.

At interview Mrs 'M' was well dressed, and gave information in a confident, concise and relevant manner. She was not unduly anxious and there was no evidence of depression or psychosis. Her mood was low but sleep and appetite were unimpaired. She was of above-average intelligence, grasped the treatment rationale well and was anxious to enter therapy.

Treatment of the irritable bowel syndrome will be by bowel retraining, to promote regular bowel movement unassociated with anxiety. In addition, specific muscle relaxation during stressful periods will be employed as a coping tactic. This will be carried out as an outpatient on an individual basis, and will include homework by the patient, in which the husband will act as co-therapist.

We will write again at completion of treatment.

Yours sincerely,

Nurse Therapist   Senior Registrar
Psychological Treatment Unit

Specimen Letter: Irritable Bowel Syndrome—Discharge

16.1.84

Dear Dr 'L',

Re. Mrs 'M'   d.o.b. 2.11.56   Hosp. No. . . . .

Address . . . .

Further to my letter of 26.7.83 this patient has been discharged from treatment. Her main problem had been pain, distension and contractions of the bowel, with associated infrequent and irregular bowel motions. The patient had 14 sessions of treatment by bowel retraining and exposure in fantasy to cues which provoked attacks, with tuition in coping tactics and self-monitoring. She became able to induce the pain at will in the clinic and then control it until it passed away. Treatment was initially slow, in order to allow monitoring of each separate component, but afterwards proceeded swiftly, with the patient being extremely compliant and performing homework tasks diligently.

The irregular bowel movements resolved within the first few weeks of treatment and have not returned, and she now has bowel movements daily. The pain and bowel contractions proved more difficult and, initially, rose in frequency following her taking a more stressful job, but are now greatly improved. Attacks currently occur less than once a month and are short—she can terminate them within 10 minutes. As a result, she now works better, socialises more and is happier at home; she has also resumed sexual activity with her husband.

One problem remains: Mrs 'M' is still upset by the abdominal distension which remains. She feels slightly unattractive and still has a sensation of fullness but thinks these are a comparatively minor issue now that she is pain free.

We will write again after seeing her for 1-month follow-up.

Yours sincerely,

Nurse Therapist
Psychological Treatment Unit

Specimen Letter: Irritable Bowel Syndrome—1-month Follow-up

28.2.84

Dear Dr 'L',

Re. Mrs 'M'   d.o.b. 2.11.56   Hosp. No. . . . .

Address . . . .

Further to my letter of 16.1.84 this patient was seen for 1-month follow-up. She remains improved in her pain, distension and contractions of bowel and has consolidated her gains. Since our last meeting she has had only one pain episode, of short duration, and was able to terminate this after 5 minutes. Frequency of bowel motions remains daily and bowel distension is unchanged.

We will write to you again after seeing her for 3-month follow-up.

Yours sincerely,

Nurse Therapist
Psychological Treatment Unit

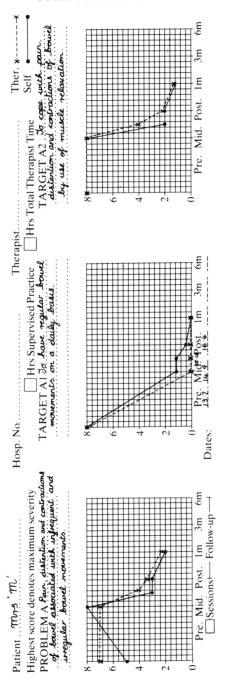

Patient *Mrs 'M'* .... Hosp. No .... Therapist ....

Highest score denotes maximum severity

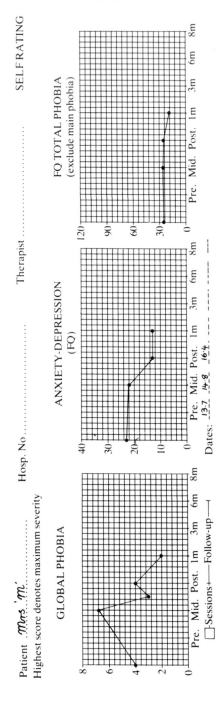

SELF RATING

FQ TOTAL PHOBIA
(exclude main phobia)

120 90 60 30 0

Pre. Mid. Post. 1m 3m 6m 8m

ANXIETY-DEPRESSION
(FQ)

40 30 20 10 0

Pre. Mid. Post. 1m 3m 6m 8m

Dates: _13·7_ _14·8_ _16·4_ ___ ___ ___ ___

GLOBAL PHOBIA

8 6 4 2 0

Pre. Mid. Post. 1m 3m 6m 8m

□—Sessions┤ ——Follow-up——┤

444

3.4.11 Specimen Letter: Unsuitable because Problem Unpredictable and Causes no Handicap

21.6.83

Dear Dr ....

Re. Mr 'N'   d.o.b. 5.5.25   Hosp. No........

Address ....

Thank you for referring this patient, whom we saw yesterday for possible behavioural psychotherapy. Such treatment is not indicated at present since his problem follows no predictable pattern and causes no avoidance or handicap, and he copes effectively with it.

His main problem is dislike of crowded places such as supermarkets and railway stations, because he may be jostled or touched by women. After this happens he has difficulty in urinating the next time he wishes to do so—even if this is many hours after; he gets a dull ache in his testicles while urinating and a broken stream of urine. He voids his bladder in 1 or 2 minutes after which the pain disappears. He cannot remember the last time this problem arose and is often free of it for months. He copes by imagining a relaxing scene before entering a crowd. In the past he was treated by hypnotherapy and systematic desensitization, with poor results.

He had an enlarged prostate which required surgery on three occasions, the last time being in 1981. This is not currently a problem.

There is no history of other psychiatric problems. At interview he was a well dressed, intelligent man who gave information readily but circumstantially. There was no sign of depression or psychosis. He was well oriented to time, place and person and his memory was satisfactory.

We agreed that his present coping mechanisms were adequate and he was encouraged to continue with them. No further appointment was made.

Yours sincerely,

Nurse Therapist   Senior Registrar
Professor Marks' Unit

3.4.12 Specimen Letter: Unsuitable due to Psychosis

22.5.83

Dear Dr 'Q' . . . .

Re. Ms 'P'   d.o.b. 5.5.45   Hosp. No. . . . .

Address . . . .

Thank you for referring this patient, who was seen yesterday on behalf of Professor Marks for possible behavioural psychotherapy. She defined her main problems as intrusive thoughts of 10-years' duration, with some accompanying rituals. She had religious doubts, thoughts that her creativity was being taken away by an external agent and preoccupation with a variety of numbers. These thoughts were sometimes accompanied by ritualistic washing and changing of clothes.

At interview she was a casually dressed, obese, shy woman who gave information incoherently and appeared suspicious throughout. Her eye contact was poor and affect was inappropriate at times. She refused to answer some questions, was extremely slow to answer others and complained of thought-blocking. There was additional evidence of psychosis as she had ideas of reference and persecution, olfactory and tactile hallucinations and passivity phenomena (externally directed acts and thoughts). There was no evidence of depression though she has had short episodes of severe depression once or twice a year. She was well oriented and memory was normal, but her concentration was extremely poor.

This lady's problem is not amenable to behavioural psychotherapy and in any case her poor concentration would make it difficult for her to carry out treatment tasks. She should be assessed by a local psychiatrist; this might be done as a day patient. In 1979 she saw a consultant psychiatrist, Dr 'X', to whom we spoke today and whose unit would be prepared to reassess her. However, she is reluctant to see a psychiatrist, as she does not feel that she has a psychological problem.

Should her psychosis remit we would be willing to reassess her if the obsessive-compulsive features persist.

Yours sincerely,

Nurse Therapist   Senior Registrar
Professor Marks' Unit

3.5 DATA CODING

3.5.1 General
It is important that data collected from each patient seen in the unit is transferred to data coding sheets. This will facilitate entry of the data into a computer to allow analysis.

3.5.2 Analyses
There are a number of analyses which can be done but the important ones are: (*a*) To check on 'quality control', i.e. we need to be sure that the effectiveness of treatment given to patients is maintained at the highest level possible. (*b*) It allows us to answer important questions about the frequency of various problems and about prognostic indicators.

3.5.3 Accuracy
It is important that these sheets are completed accurately and clearly in *pencil* so that errors can be corrected readily.

If you have any uncertainty about how to complete any patient's data, please don't hesitate to ask one of the course supervisors.

4. *Appendices*

4.1 GENERAL ADMINISTRATIVE ISSUES

4.1.1 General

Names, dates, reference numbers, addresses, phone numbers, appointments, specified agreements, times of calls, etc. are the hard data on which the service depends.

Generally: CHECK IT
           WRITE IT DOWN
           GIVE A COPY TO OTHERS INVOLVED

4.1.2 Office Mechanics

The essential components are:

Display Board (office)

All cases being processed by the unit, including the waiting list, should be shown on the case flow board. Before screening, the name is put up by the secretary. After screening it is the therapist's responsibility for passing across to the appropriate sections of the board. The sections are:

1. Waiting list—kept by the secretary.
2. Screening—names of patients to be screened.
3. Therapist—suitable patients, who have been screened and assessed and are currently in treatment, discharged or being followed-up.
4. No show—if the patient does not attend screening, it is the delegated therapist's responsibility to move the name to this section.
5. Unsuitable—if patient is considered unsuitable for treatment at screening.

N.B. The patient's name, etc. should only be shown in one section of the display board.

Case flow board (Secretary's Office)—to be maintained by trainees using numbers only. Example ....

'Post screen allocation'	— Record this each time you are allocated a patient to assess (not screening).
'DNA'	— Did not attend for assessment (may have attended for screening).
'Assessed'	— Record when full assessment has been completed.
'Unsuitable'	— If unsuitable for treatment after assessment, even if considered suitable at screening.
'Refused'	— Lapsed following assessment, i.e. failed to attend session 2.

N.B. Assessment should be considered as session 1. Patients must have been suitable before they can be considered as having refused or dropped out.

'Suitable'	— Fulfils all selection criteria at assessment *see* pp. 3–5)
'Current treatment'	— Allocate suitable patients according to clinical category, e.g. specific phobia, agoraphobia. Use primary problem if more than one is present. This section should not include 'refusers' or 'drop outs'.
'Discharged'	— After an adequate trial of treatment which usually takes 6 or more sessions—but sometimes takes less.
'Dropped out'	— Discontinued treatment after an inadequate trial, between sessions 3–6.

N.B. Patients should not be shown in more than one category.

From the example of the figures for Therapist 1 (T.1) (right) we can see that 2 patients did not attend for assessment and 3 are awaiting assessment (19 − [2 + 14]), 6 are in current treatment, 2 have been discharged after an adequate trial, and none have dropped out.

Therapist	Post screen allocation	DNA	Assessed	Disposal			Current Treatment						Discharged						Drop out	Completed adequate trial & 1 MFU
				Un-suit	Suit	Ref use	Spe-cif	Soc	Ag	Ob	Sex	Mis-cel	Spe-cif	Soc	Ag	Ob	Sex	Mis-cel		
T.1	19	2	14	4	10	2	1	1	2	1	1	0	1	0	1	0	0	0	0	0

Message book—all general messages, both internal and from outsiders, are transmitted via this book, not via the secretary; all staff should refer to the book at least three times daily.

Appointments book—appointments, bookings of facilities (e.g. common room, TV equipment, etc.) and visits of outside professionals, etc., must be booked in advance.

Filing cabinets—stick to alphabetical order in appropriate sections of appropriate cabinets. When removing medical notes a tracer card must be left in their place.

Stocks (of forms, etc.)—therapists should let the secretary know when any stock is running low.

Priority trays—the secretary's 'in-tray' is divided into high, medium, or low priorities: use these realistically (e.g. routine correspondence is not high priority, unless long delayed).

The secretary—is overworked and also human. She responds to flattery only if it is linked with practical co-operation, such as:
1. Realistic use of the priority trays.
2. Clear handwriting and dictaphoning.
3. Prior warning and discussion of unusual items.
4. Keeping out of the office, except for business.
5. Not using her as a substitute message book.
6. Filing of all reference folders and hospital notes in strictly alphabetical order.

4.1.3 Case Load Planning

Short-term factors
The following factors determine the limits within which work rate can vary. Each treatment package should generally be 'short and fat', rather than 'long and thin', e.g. an investment of 5–20 therapist hours over a total of 2–10 weeks. The justification for this is both clinical and administrative.

Given that each case demands 2–6 therapist hours per week and that a maximum of 20 hours per week is available for casework (as distinct from other forms of training), the case-load should be in the range of 3–7 at any one time during training.

Long-term factors

The following must be considered when periodically adjusting case-load:

1. A minimum of 12 patients must complete an adequate trial of treatment within 12 months.
2. Those cases have to be distributed across the 6 'main problem' categories (agoraphobia, social or specific phobia, obsessive-compulsive disorder, habit disorder, sexual dysfunction or deviation, other), such that in at least 5 categories, 2 cases have been completed.
3. The effective therapy period is 9 rather than 12 months, if the introductory period, the holidays, and the tidying up period are discounted.
4. Up to 30% of cases taken on will fail to complete an adequate trial. This may be 50% in the sex categories; and thus at least 17 cases should have started treatment by month nine.

Monitoring work flow (case flow chart)

Each therapist is expected to actively monitor his clinical work rate and match it with requirements—e.g. by pressing for cases in categories which are falling short. The case flow chart is designed to facilitate an ongoing record of each patient referred to the unit and their progress from referral to follow-up. It also serves as a checklist for relevant correspondence, which will assist both therapist and supervisor in ensuring that this is carried out promptly. The chart should be kept up to date and be presented at each formal personal supervision.

Guidelines for using the case flow chart

1. The name of each patient allocated to a therapist should appear on the chart, including those who do not attend for screening interview.
2. Each patient should be given a diagnosis whether suitable or unsuitable. For those patients found unsuitable at screening or assessment, a reason must be given for their unsuitability, e.g. severe depression, psychosis, no clear behavioural goals, declined treatment, etc. Patients who fail to attend for a screening interview should be entered as DNA (did not attend).
3. The date the screening interview took place should be entered, and the date on which the screening letter is signed by the therapist. The same applies for assessment, discharge and follow-ups. The date that a case summary is completed (at 1-month follow-up) and typed, should be entered in the appropriate column.
4. The number of treatment sessions for each patient must be continually updated for supervision. Assessment is recorded as Session

PATIENT	DIAGNOSIS (Give reason if unsuitable)	DATE SCREENED	DATE LETTER	DATE ASSESSED	DATE LETTER	SESSION No or STATUS i.e. DO; DNA; REF.	DATE DISCHARGED	DATE LETTER	DATE 1 MONTH FU	DATE SUMMARY	DATE 6 MONTHS FU	DATE LETTER	DATE 1 YEAR FU	DATE LETTER	COMMENTS

CASE FLOW-CHART (spanning: DATE ASSESSED, DATE LETTER, SESSION No or STATUS, DATE DISCHARGED, DATE LETTER, DATE 1 MONTH FU)

Number 1. Use the same column for patients who do not complete treatment and enter as appropriate:

DO (drop out) = discontinued treatment after an inadequate trial, from Session 3 up to Session 5.

N.B. Six sessions are regarded as an adequate trial; however, it is possible that less than 6 sessions can be an adequate trial determined by outcome.

REF (refused) = discontinued treatment following assessment Session 1.

DNA = did not attend for assessment.

5. The Comments column is for remarks relevant to each case, e.g. cancelled sessions, fails to respond to correspondence, follow-up ratings by post, relapsed at follow-up, booster sessions given, etc.

4.2 FILE CONTENTS

Case Folder

Contents (in order):

1. Data summary sheet (followed by case summary when completed after 1-month follow-up).
2. Copies of all letters to and from referral source, other agencies or the patient, most recent letter first.
3. Copy of referral letter.
4. Graphs of outcome.
5. Assessment write-up.
6. Progress record.
7. Clinical measures—Problems and targets

General measures (FQ, work, home, social, private)

Special measures, e.g. obsessive-compulsive, sex, etc.

8. Homework records.
9. Storage: in the common room, white cabinet (locked).

Medical Folder

Must carry:

1. Copies of major letters at:
   screening
   assessment
   discharge
   follow-up, 1-month, 3-month and 6-month

2. Data Summary Sheet }
3. Case Summary   } when completed
4. Graphs
5. Storage: in the general office white cabinet while current; back to Medical Records Department after 1-month follow-up, with front sheet completed by senior registrar.

4.3 TRAINING FACILITIES

4.3.1 Library

Facilities are available at the Maudsley Hospital, Psychological Treatment Unit which include a wide range of books, journals and reprints on behavioural psychotherapy and psychiatric nursing. Books and journals can be borrowed (1–2 at a time) for 2 weeks and trainees must endeavour to return these in time for use by their colleagues. In the absence of a librarian it is the user's responsibility to keep the books, journals and articles in their respective places.

The libraries at the Institute of Psychiatry and School of Nursing offer their facilities to trainees. The former stocks books and journals on behavioural psychotherapy and related subjects. It also has facilities for computerized literature search and ordering books and reprints from other libraries.

4.3.2 Audiovisual Facilities

Video work is an important part of the training programme. The studio has been recently updated at a considerable cost. It is available for supervised interviewing and making teaching tapes for future use. The service of a part-time video technician is available. A small number of videotapes are available for teaching and clinical purposes. Strict rules apply to interviewing patients on closed circuit television (CCTV) and when recording an interview. These guidelines are given at the end of this section and should be rigidly adhered to. Borrowing of videotapes or their viewing outside the unit is not permitted.

Audio facilities

Excellent recording and playing facilities of audiotapes are available. These can be used for teaching and clinical purposes (e.g. to aid imaginal exposure in thunder phobics).

Teaching aids

The conference room has facilities for videotapes, slide and overhead projectors. Extensive use of these is encouraged whilst teaching or presenting cases. A small number of slides are available for teaching and clinical purposes.

4.3.3 Consent for Videotaping Patients

1. Before making videotapes involving patients or their relatives, interviewers are asked to acquaint themselves with the *code of practice* which is available in all studios and recording rooms.
2. In particular, they are reminded that the patient's *verbal consent* should be recorded and that special procedures should be used for patients incapable of giving consent.
3. Each tape should include at the beginning the *copyright statement* and the *code number* and should be accompanied by a duly completed *index card*.
4. Tapes involving patients who have given consent may be shown freely within the Joint Hospital and Institute to people professionally involved in the treatment of mental illness. *Any showing outside the above* category must have prior approval by the consultant in charge of the case and the chairman or vice-chairman of the Clinical Research Review Committee.

Code of Practice in Making and Using Videotaped or Filmed Records of Patients

Introduction
This memorandum is concerned with videotapes or films or other material (including slides) from which the patient can be identified. The main problem in making and using such records is that of confidentiality. The following proposals are put forward in order to ensure that breaches of confidentiality do not occur, but we have found it impossible to lay down rules to cover every contingency. Whenever the word 'videotape' or 'film' is used, it should be interpreted to include slides or other similar material, when relevant.

Making videotapes or films
1. The copyright* of a videotaped or filmed record is held by the employing body of the person responsible for making it. Every such record should be given a number and entered in a register. This

---

* A tape may be owned by a body which does not hold the copyright of the material recorded on it. If the tape is returned to such a body, the content should first be erased, unless the procedure described under 'Showing Videotapes and Films (2)' has been followed.

number, the name of the responsible person and the organization holding the copyright should be recorded on the tape or film, together with the statement, 'This recording may not be shown or reproduced without permission'. The same information should be written on the record container.

2. The responsible person will ordinarily be a consultant of the Joint Hospital or senior member of the Institute staff (senior lecturer or above). It is the responsibility of this person to ensure that anyone on his staff to whom he delegates authority to make videotaped or filmed records is aware of the necessity to adhere to this code of practice.

3. When the responsible person is not himself the patient's consultant, the permission of the consultant or his deputy should be obtained before making the record.

4. The patient's consent to the procedure should be obtained whenever possible before recording begins and again at the beginning of the recording session. A suitable form of wording for the latter would be, 'As I explained to you, this interview is being recorded on videotape and may be used for purposes of education or research. Strict confidentiality will always be observed. Is that all right with you?' It should be remembered that no form of consent gives legal protection if harm comes to the patient as a result of misuse of the recording.

5. Any conditions laid down by the consultant, the person responsible for making the tape, or the patient as to the way in which the videotape or film is to be used must be accepted and written on the record container and spool.

6. In the case of children under 16 years of age, written permission must be obtained from parent or guardian. The person responsible for making the tape should decide whether to use the form of words suggested above, having regard to the patient's age and other circumstances.

7. Unless there is some special reason for doing otherwise, the interviewer should avoid addressing the patient by name during the recording.

8. If the patient is unable to give or withhold consent, the decision as to whether to proceed is the responsibility of the person making the recording, and the fact that consent has not positively been obtained, should be noted on the spool and container.

9. If the patient refuses consent, no record should be made. If the patient withdraws consent during the recording or subsequently, the interview should be discontinued and the tape erased or film destroyed.

Showing videotapes and films

1. Once permission has been given by the patient and the relevant consultant for making the recording, no further permission should be

required for showing that videotape within the Joint Hospital or Institute to people professionally involved in the treatment of mental illness. The phrase 'professionally involved' should be taken to include doctors, nurses, psychologists, social workers, occupational therapists and those undergoing training for these professions.

2. Anyone wishing (*a*) to show a videotape of a patient to a non-professional audience, (*b*) to show a videotape of a patient to any audience outside the Joint Hospital or Institute, or (*c*) to give or lend a videotape of a patient to any person outside the Joint Hospital or Institute must obtain written permission from whichever person or body holds the copyright of the tape before he does so. This procedure should also be followed in the case of any query.

The written permission of the chairman or vice-chairman of the Clinical Research Committee and the consultant responsible for the patient would also be required. The decision would depend upon an assessment of:

a. The content of the videotape (e.g. whether potentially embarrassing personal information was revealed).

b. The composition of the intended audience.

c. The likelihood of the patient being recognized by members of the audience.

d. The reason for showing the tape (teaching or research) and the feasibility of achieving the same object by other means.

e. The likelihood of the persons receiving the tape abiding by this code of practice.

Custody of videotapes and films

1. One or more persons nominated by the Clinical Research Committee will be responsible for the registration and safe custody of videotapes and films including their return after being loaned out.

2. When not in use tapes should invariably be kept in locked cabinets and access should be limited to authorized people only.

3. All tapes must be erased before disposal.

Tapes and films which have already been made

The suggested rules concerning showing, storing and disposal of tapes should also be applied to tapes and films which have already been made by Institute and Hospital staff. This means that they should be numbered and registered. It will be assumed that the patient's consent has been obtained, although in many cases it will not be recorded on the document.

Procedures

1. *Standard consent form for recording interviews:* These must be obtained from the responsible supervisor and completed prior to each recording. A serial number will be given with the form. The form is in triplicate and is to be distributed as follows: copy 1 to be kept with the film; copy 2 to be given to the supervisor; copy 3 to go into the patient's medical records.

2. *Other documentation:* It has also been agreed that in future video-taped material should be documented as follows:

a. Tapes to be numbered rather than named to preserve patients' confidentiality. Each cassette will be numbered and each recording on each cassette will be numbered in addition.

b. A separate card index is to be kept giving the tape number, the patient's hospital number, the patient's name and consultant, the interviewer, the other people present at the interview, the diagnosis and the date the tape was made. This will be kept by the responsible supervisor.

3. *Making the videotape:* Those wishing to make videotape recordings are referred to the attached code of practice. It is suggested that on hospital sites technicians will have requested and be able to advise on the operation of the equipment but will not normally be available to operate equipment. Each department or board having videotaping equipment should take steps to nominate certain members of staff who will operate the equipment. Such a system will help to ensure that the equipment is not misused.

4. *Consent form for the showing of sexually explicit material* (*see* p. 129): Where the viewing of sexually explicit material, whether film slides or pictures, is essential in the treatment of patients, the patient's signed consent must be first gained (Section A). The therapist must sign to confirm that an explanation of the nature and purpose of the material has been given (Section B). The form must also be signed by a supervisor. All three parts of the form must be completed before showing any sexually explicit material. The completed form should be inserted in the patient's medical records.

To be kept with Film (1)
To Institute Video Library (2)
To Patient's case notes (3)

INSTITUTE OF PSYCHIATRY/BETHLEM ROYAL HOSPITAL AND MAUDSLEY HOSPITAL
Consent Form – Recorded Interviews

I consent to the recording of an interview with me/my family* being made and kept on videotape/audiotape.

I understand that this recording may be used for purposes of assessment, teaching or research. Strict confidentiality will always be observed, and it will be seen only within the Institute of Psychiatry and the Bethlem Royal and Maudsley Hospital by professional staff or their trainees.

(Any permission for wider training to be noted here and separately signed and dated.)

Names of all those appearing (Parents also signing on behalf of a child should write 'parent') (Relatives signing on behalf of a patient unable to give consent should state relationship)	**Age** (if under 18)	**Signatures**
....	....	....
....	....	....
....	....	....
....	....	....
....	....	....

Name of Consultant ....

Name of Interviewer   **Signature of Interviewer**

....

Date ....   Serial number of tape ....

This form must be signed at the conclusion of the recording by all those who appear on the recording. In the case of young children, the parent or guardian should sign, or in the case of patients unable to give consent, their nearest relative should sign on their behalf.

The complete form should be filed in the patient's notes.

PLEASE NOTE that it is still necessary to inform the interviewees during the first part of the recording that a recording is being made, and that their written permission for its preservation will be requested at the end.

4.4 SECONDMENT

This section only applies to some therapist trainees for whom the final six months of ENB (formerly JBCNS) 650 comprises a period of secondment to centres away from the training hospital. This may be in a hospital or primary health care setting, or a mixture of both. Record keeping is essentially the same as at the parent hospital, with the addition of certain records to allow scrutiny at the training hospital.

There are also several modifications to clinical practice:
1. More liaison with outside agencies (GPs, social workers, CPNs).
2. Less freely available medical screening.
3. Less access to supervision.
4. More autonomy/wider range of referrals and referral sources.
5. Larger teaching role.
6. Less knowledge from supervisors at the seconding hospital.
7. Little contact with colleagues trained in behavioural psychotherapy.
8. Larger caseload/responsibility to provide a service.

These factors may contribute to the fact that trainees on all previous courses have found secondment a key experience in developing the skills necessary to become autonomous practitioners. In addition to the normal components of therapy, trainees find themselves involved in:
1. Setting up a new service and negotiation with nursing and medical staff to identify and consolidate the role of that service.
2. Publicizing the service, both internally and externally.
3. Administering the service—constructing waiting lists, job descriptions, recording and analysing throughput of patients.
4. Arguing the needs of the service with management.
5. Negotiating physical resources.
6. Broadening the role of nurse therapist in response to specific local needs.
7. Resisting pressure to take on patients obviously unsuitable and unresponsive to behavioural psychotherapy.

The above represents a marked difference in focus from the initial phase of training, which centres on specific therapeutic skills learned under close supervision, and is more akin to the latter stages of treatment of patients, where generalization begins and the therapist's direct involvement fades.

TREATMENT OF SEXUAL PROBLEMS
Consent to the viewing of sexual films (including videotapes)

To the Board of Governors,
The Bethlem Royal Hospital and the Maudsley Hospital

A

I/We .... and ....

of ....

hereby consent to the viewing of films depicting sexual actions.
I/We understand that this material is to help me/us with my/our treatment.

Date .... Signature .... Name ....

Signature .... Name ....

B

I confirm that I have explained the nature and purpose of the films to the above named.

Date .... Signature .... Name ....
of therapist

C

During the treatment of the above-names, the showing of:
    a) explicit heterosexual films
    b) other material (specify) ....
will be required.

Date .... Signature .... Name ....
Consultant/deputy

Note: A, B, and C must be completed in all cases before films are shown.

This Consent Form is to be firmly affixed to the patient's records

Index

adaptation, 1
administrative issues, 116–21
agoraphobia, 2, 3, 8, 13, 24–5, 28, 30, 32–5, 67–9
   interview for, 67–9
alcohol dependence, 18
anger control, 4, 14, 70–1
anorgasmia, 2, 4, 11, 14
anxiety, 8, 13, 14
anxiety-depression questionnaire, 47, 60
assessment, 22–6
aversion, 6, 8, 10, 14, 96–102
avoidance, 8

behavioural family therapy, 5
behavioural goals, 4, 5, 11, 18, 26, 27
behavioural guidance, 9, 19
behavioural medicine, 5, 15
behavioural psychotherapy
   definition, 1
   indications, 3–5, 18
   principles, 5
bell and pad treatment, 9
bowel retraining, 109–10
breathing exercises for stammering, see stammering
bulimia, 4, 14, 69–70

cardiac phobia, 77–81
case flow board, 117
case reporting, 36
case summary, 32–4
caseboard planning, 118–20
children's problems, 15, 16
clinical management, 17–42
closed circuit television (CCTV), 19, 123–7
coaching, 2
cognitive rehearsal, 1
cognitive therapy, 6–8, 13
compliance, 7
compulsion checklist, 49, 60
compulsive rituals, 1–4, 11, 74–7, 91–3
confidentiality, 40
consent of patient, 40
contextual therapy, 1
contingency management, 2, 6
contracting, 6, 9

covert sensitization, 6, 9, 14, 72, 99, 102
cue exposure, 6, 9

data coding, 115
data summary sheet, 44
decision flow in treatment, 41
dental anxiety measure, 50
depression, 4, 8, 13, 14, 18
desensitization, 1, 11
diary for homework, 56
discharge of patient, 29–30
display board, 116

eating disorders (see also bulimia), 14, 101–5
ectopic heart beats, phobia of, 77–81
enuresis, 3, 4, 9, 10
erectile failure, 1, 11
ethics, 40
exhibitionism, 4
exposure, 1–4, 6, 9, 15, 25, 75
extinction, 1, 9

fading, 9
fear questionnaire, 37, 47, 60
feedback, 5, 10
fetishism, 4, 98–101
file contents, 121–2
flooding, 1, 6, 9, 10
follow-up, 2, 31–2

gagging, 9
generalization, 9
graphs, 40, 59–65, 81–3, 88–90, 111–12
grief, 9, 13, 14
guided fantasy, 6, 9
guided mourning, 9, 14

habit control, 6, 9
habit disorders, 4, 71
habit reversal, 9
habituation, 1, 2
hairpulling, 4
hierarchy, 9
homework, 2–4, 7, 11, 19, 26

homework diary, 56
  for sexual problems, 57

illness behaviour, 4
implosion, 9, 10
injection phobia, 84–90
instrumental conditioning, *see* operant conditioning, shaping
interviewing
  assessment, 23
  schedules, 66–72
  screening, 18
irritable bowel syndrome, 5, 108–12

latency, 10
leisure adjustment measure, 46, 61
letters to referral agents, 19–21, 30, 34–5, 74–114

marital problems, 14
masochism, 4
massed practice, 10
measures, 5, 12, 37–40, 43–115
modelling, 1, 2, 6, 10, 11

nightmares, 4, 10, 14

obesity, 14
obsessive-compulsive measures, 48–60
obsessive-compulsive disorder, 1–4, 8, 11–13, 27, 37, 39, 91–3
operant conditioning, 1, 10
orgasmic reconditioning, 10

pacing, 2, 6
paedophilia, 4
pain, 15
paradoxical intention, 1, 10
phobias, 1–3, 10, 13, 27
  being alone, 77–81
  contamination, 91–3
  ectopic heart beats, 77–81
  injections, 84–90
  social, 3, 10, 21, 24–5, 30, 32–5
  spiders, 74–7
premature ejaculation, 2, 11, 72, 96–8
principles of behavioural management, 5
problem measures, 33, 37, 38, 45, 59
prompting, 2, 6
psychosis, 4, 18

rehearsal relief, 10, 14
reinforced practice, 10
relaxation, 2, 10
response cost, 2, 11
response prevention, 6, 11, 15
restructuring, 10
rituals, 1, 2, 11, 74–7, 91–3
roleplay, 6

satiation, 11
schizophrenia, 5, 15
screening, 17–21
secondment, 128
self regulation, 2, 4, 6, 10, 11, 13, 15, 34–5
self-help manual, 2, 13
semantic differential measures, 53–5
sensitization, 2
  covert, 6
sexual activity measure
  conventional, 51, 62
  unconventional, 52, 63
sexual attitudes, 39, 53–5, 64, 65
sexual deviation, 1, 3, 4, 10, 37, 39, 71, 98–101
sexual dysfunction, 2–4, 14, 37, 71, 73, 96–8
sexual homework diary, 57
sexual problems, interview for, 71
sexual skills training, 2, 3, 6, 11
shaping, 1, 6, 10, 11
shyness, 3, 13
sleep management, 16
social adjustment measures, 46, 61
social phobias, 3, 10, 21, 24–5, 30, 32–5
social situations questionnaire, 58
social skills problems, 1, 3, 4, 6, 8, 15, 93–6
social skills training, 2, 3, 11, 13, 21, 73, 74, 93–6
spider phobia, 74–7
squeeze technique for premature ejaculation, 96–8
stammering, 3, 4, 9, 10, 14, 37, 105–8
strategy of treatment, 41–2
successive approximation, 10
suitability for behavioural treatment, 18
symptom substitution, 1

target behaviour, 12, 33, 37–9, 45, 59
thought stopping, 6, 12
tics, 1, 9, 14
time out, 12
token economy, 10, 12, 15

torticollis, 9
training facilities, 122
treatment session, 26–8

vaginismus, 4, 11

work adjustment measure, 46, 61